Environmental Reporting and Recordkeeping Requirements

 Government Institutes, Inc.

PUBLISHER'S NOTE

This publication is designed to provide accurate and authoritative information with regard to the subject matter covered. It is sold with the understanding that the publisher and authors are not engaged in rendering legal, accounting or other professional service. If legal advice or other expert assistance is required, the services of a competent professional person should be sought.--Adapted from a Declaration of Principles jointly adopted by a Committee of the American Bar Association and a Committee of Publishers.

Publication of this book does not signify that the contents necessarily reflect the views or policies of Government Institutes, Inc.

ALL RIGHTS RESERVED

No part of this publication may be reproduced, stored in a retrieval system, or transmitted in any form or by any means, electronic, mechanical, photocopying, recording or otherwise, without the prior written permission of the publisher.

KF
3775
.E55
1990

July 1990
Second Printing, December 1990

Published by
Government Institutes, Inc.
966 Hungerford Drive, #24
Rockville, Maryland 20850-1714
U.S.A.

Proceedings of the
Environmental Reporting and Recordkeeping Requirements Course
March 1990

Copyright © 1990 by Government Institutes, Inc.
ISBN: 0-86587-216-3

Library of Congress Catalog Card Number: 90-82901
Printed and bound in the United States of America.

TABLE OF CONTENTS

About the Authors iv

Chapter 1 Introduction to Environmental Reporting
and Recordkeeping Requirements 1
Theodore W. Firetog
Environmental Counsel
Shea & Gould

Chapter 2 Clean Water Act. 5
Robert L. Collings
Partner
Morgan, Lewis & Bockius

Chapter 3 Resource Conservation and Recovery Act (RCRA). 38
David R. Case
General Counsel
Hazardous Waste Treatment Council

Chapter 4 Clean Air Act 51
G. Vinson Hellwig
Vice President, Air Division
TRC Environmental Consultants, Inc.

Chapter 5 Comprehensive Environmental Response, Compensation and Liability Act
(CERCLA/Superfund) 64
Theodore W. Firetog
Environmental Counsel
Shea & Gould

Chapter 6 Occupational Safety and Health Act (OSHA) 76
Robert D. Moran
Attorney
Cooter & Gell

Chapter 7 SARA Title III/Emergency Planning
and Community Right-to-Know Act (EPCRA). 121
J. Gordon Arbuckle, Partner
Timothy A. Vanderver, Jr., Partner
Paul A. J. Wilson, Attorney
Patton, Boggs & Blow

Chapter 8 Toxic Substances Control Act (TSCA) 155
Charles A. O'Connor, III, Partner
Thomas B. Johnston, Associate

Appendix A Guide to Record Retention Requirements
for 40 CFR (EPA) and 29 CFR (OSHA) 125

ABOUT THE AUTHORS

J. Gordon Arbuckle
Partner
Patton, Boggs & Blow
Washington, D.C.

Mr. Arbuckle is a partner who has practiced environmental and natural resources law for over 20 years. He has written and spoken extensively on the Clean Water Act, TSCA, hazardous waste, biotechnology, and other areas of environmental law.

Among his major endeavors, he represented the consortium of businesses that formed the Louisiana Offshore Oil Port in the legislative and administrative efforts necessary to obtain governmental approval, complete construction and continue operations.

He also participated in the successful efforts to obtain authorization to construct the Alaska natural gas pipeline and was lead counsel for the consortium which obtained the first deep seabed mining license. He played a major role in developing and securing enactment of the Deepwater Ports Act and the Deep Seabed Hard Mineral Resources Act, as well as certain provisions of other environmental laws, including RCRA, the Clean Water Act and CERCLA.

Mr. Arbuckle has represented clients in petitions filed under RCRA and in hearings on Clean Water Act permits. He also been extensively involved in Superfund litigation and is among the authors of "The 1986 Superfund Amendments: An Analysis of Selected Provisions," _Toxics Law Rep._ (Feb. 4, 1987). He is an author of the Government Institutes' _Environmental Law Handbook_, a text widely read by environmental practitioners, and of the _Emergency Planning and Community Right-to-Know Handbook_. He has authored several articles concerning criminal liability under environmental statutes as well.

David R. Case
General Counsel
Hazardous Waste Treatment Council
Washington, D.C.

Mr. Case is the General Counsel of the Hazardous Waste Treatment Council (HWTC), a national trade association of more than 60 commercial firms engaged in high-technology treatment, remedial site cleanups, and supporting equipment manufacturers.

Mr. Case currently represents the HWTC in litigation in the U.S. court of Appeals in Washington, D.C. concerning the U.S. EPA's regulatory program for hazardous waste management under RCRA. He worked on the HWTC rulemaking before the EPA to establish a new RCRa permit program for mobile treatment units. He is also responsible for HWTC's program for monitoring the selection of cleanup remedies that utilize alternative treatment technologies at Superfund sites, as required by SARA. HWTC and a coalition of environmental groups recently published the first comprehensive report on EPA's cleanup decisions under SARA entitled "Right Train, Wrong Track."

Mr. Case is a frequent lecturer and author on environmental law. He received a B.A. from Amherst College (magna cum laude), an LL.B from Cambridge University, England, and a J.D. from the University of Michigan Law School. He is a member of the District of Columbia and Michigan Bars.

Robert L. Collings
Partner
Morgan, Lewis & Bockius
2000 One Logan Square
Philadelphia, Pennsylvania 19103-6993

Mr. Collings is a Partner in the Philadelphia law firm of Morgan, Lewis & Bockius's Government Regulation Section. His practice involves a wide variety of environmental law matters, including citizen suit defense, defense of government enforcement actions, permit proceedings, Superfund claims, and counseling in business transactions.

Prior to joining the firm, Mr. Collings was Associate Regional Counsel and Branch Chief of the U.S. EPA's Region III Office with responsibilities in all of the major environmental programs.

He is listed in <u>The Best Lawyers in America</u> (Woodward/White, inc. 1989) for his work in the environmental law. He is so the author of many articles on environmental laws and regulations, inlcuding Superfund, infectious waste, lender liability, and other topics.

Mr. Collings is a graduate of Harvard College (A.B. 1972) and Boston College Law School (J.D. 1977).

Theodore W. Firetog
Environmental Counsel
Shea & Gould
1251 Avenue of the Americas
New York, New York 10020-1193

Mr. Firetog (Program Chairman) is currently Environmental Counsel with Shea & Gould, where he provides a wide range of legal counseling on all aspects of environmental law, including regulatory compliance, indoor pollution, and hazardous waste matters.

His environmental law expertise has been built on the practical experience that he received as a Staff Attorney with the Environmental Law Institute and as an Attorney with the U.S. Environmental Protection Agency in Washington, D.C.

Following EPA, and prior to joining Shea & gould, he was associated with Rivkin, Radler, Dunne & Bayh, where he was engaged in a broad spectrum of environmental litigation.

Mr. Firetog received his B.S. degree in Natural Resources and his M.S. degree in Natural Resource Policy and Management from the University of Michigan. He obtained his Juris Doctor degree from the State University of New York at Buffalo, where he was also a Sea Grant Law Fellow.

G. Vinson Hellwig
Vice President and Manager, Air Division
TRC Environmental Consultants, Inc.
800 Connecticut Blvd.
East Hartford, CT 06108

Mr. Hellwig is Manager of TRC's Air Division where his responsibilities include project management of permitting and air engineering projects. Mr. Hellwig also serves as Senior Project Manager on a number of air projects including emissions measurements, permitting, SARA Title III inventories, and control equipment engineering.

Prior to joining TRC, Mr. Hellwig was Senior Environmental Engineer in the North Carolina office of a major environmental engineering firm. Responsibilities included regulatory analyses and air program development as well as data systems management. He was the Technical Advisor for the BACT/LAER Information System and for an update of the SIP Tracking System. Mr. Hellwig managed projects dealing with SIP development, NSPS, and NESHAP. He authored the Good Practices Manual for the Delegation of NSPS and NESHAP Regulations. While working for U. S. EPA Region IV, Air Enforcement Branch, Mr. Hellwig was responsible for coordinating reporting requirements for compliance with the eight states in Region IV, conducting inspections of facilities in a variety of industries, developing compliance strategies for specific types of sources, providing guidance to State and local air pollution agencies, and interpreting EPA policy. Mr. Hellwig was also previously employed by the Alabama Air Pollution Control Commission and the Ampex Corporation Magnetic Tape Division.

Mr. Hellwig received his B.S. degree in Chemistry from Shorter College, and completed his Graduate course work for a M.S. degree in Chemistry at Clemson and Auburn Universities.

Thomas B. Johnston
Attorney
McKenna, Conner & Cuneo
1575 Eye Street, N.W.
Washington, D.C. 20005

Mr. Johnston practices in the area of environmental law, specializing in the regulation of toxic substances and pesticides. He is an Associate Professorial Lecturer at Law at the George Washington University National Law Center, where he lectures on the Toxic Substances Control Act (TSCA) and related topics, and has lectured elsewhere on TSCA and the laws regulating pesticides. He has co-authored several publications on TSCA, including the TSCA Handbook, Second Edition.

Mr. Johnston is a member of the Pennsylvania and District of Columbia Bars and the American and Pennsylvania Bar Associations. He is active in the D.C. Bar Sections on Administrative Law and Agency Practice and Environmental, Energy, and Natural Resources and is a member of the Society for Risk Assessment. From 1979-86 Mr. Johnston served as a fisheries and wildlife biologist with U.S. Fish and Wildlife Service and the U.S. Environmental Protection Agency's Office of Pesticide Programs.

Mr. Johnston received his B.S. degree in Biology from the College of William and Mary, and his M.S. degree in Zoology (emphasis Aquatic Ecology). He received his J.D. from George Washington University.

Charles A. O'Connor, III
Attorney
McKenna, Conner & Cuneo
Washington, D.C.

Mr. O'Connor joined the Washington, D.C law firm of McKenna, Conner & Cuneo in 1970, and became a partner in 1974. He is also Chairman of the firm's Environmental, Food and Drug, and Trade Association Department.

Mr O'Connor specializes in environmental law with emphasis on pesticides and toxic substances. He represents chemical manufacturers and their trade associations in federal and state litigation and regulatory proceedings on health and environmental issues. In addition, he counsels industry joint testing ventures under FIFRA and TSCA, and is general counsel to the Chemical Specialties Manufacturers Association.

He is co-author of the <u>TSCA Handbook</u>; the <u>Pesticide Regulation Handbook</u> (Rev. Ed.); "Industry's Role in Cancer Research: Anticipating Regulatory Problems" (<u>Regulatory Toxicology and Pharmacology</u>, 1981) and the <u>Consumer Product Safety Handbook</u> (Government Institutes, Inc. 1973).

Prior to joining McKenna, Cuneo & Connor, He began his legal career as an officer in the United States Navy Judge Advocate General's Corps (1967-1969).

Mr. O'Connor received his A.B. <u>cum laude</u> at Harvard University and his J.D. at Georgetown University.

Timothy A. Vanderver, Jr.
Partner
Patton, Boggs & Blow
2550 M Street, N.W., 8th Floor
Washington, DC 20037

Mr. Vanderver is a partner in the law firm of Patton, Boggs & Blow located in Washington, DC. He is involved primarily in environmental concerns, including the National Environmental Policy Act, water pollution law, air pollution law, solid waste disposal and land development. His previous experience includes government service in senior positions of responsibility where he specialized in environmental law with the Department of Interior and the Department of Housing and Urban Development.

After receiving his Undergraduate degree from Washington & Lee University, he obtained his law degree from Harvard. He was selected as a Rhodes Scholar and received his Masters in law from Oxford University.

Paul A. J. Wilson
Attorney
Patton, Boggs & Blow
2500 M Street, N.W., 8th Floor
Washington, DC 20037

Mr. Wilson is an associate with Patton, Boggs & Blow in Washington, DC, where he has practiced in the areas of Superfund, EPCRA, RCRA, and TSCA, as well as other major environmental laws.

From 1975 through 1980, Mr. Wilson worked at the U.S. Environmental Protection Agency in the pesticides, mobil source air pollution control and toxic substances programs. While with the EPA, he was twice awarded the agency's Bronze Medal for commendable service.

Mr. Wilson holds a B.A. from the University of Massachusetts, an M.A. from Northwestern University, and a J.D. from the University of Virginia, where he was executive editor of the **Virginia Journal of Natural Resources Law.**

INTRODUCTION

THEODORE W. FIRETOG
Environmental Counsel
Shea & Gould

Until recently, environmental reporting and recordkeeping requirements have not been in the forefront of the concerns of the regulated community. Compliance with environmental laws has focused on controlling the discharge of pollutants and contaminants into the environment. As each new environmental law became enacted, the regulated community quickly focused on the subject matter at hand and braced itself for the new wave of regulations and litigation which would inevitably occur.

Indeed, the regulations, litigation, and reactions of the regulated community, which ensued as a result of the enactment of numerous federal environmental statutes over the past 20 years, have often been described as a series of waves similar to the waves of any ocean or major body of water. Each wave of the environmental movement has had its crests or moving swells, undulating with the concerns of the nation as a whole.

There was, for example, the wave of regulations and litigation following enactment of the Occupational Safety and Health Act of 1970 (29 U.S.C. §651 et seq.). There was the wave of regulations, litigation, and enforcement resulting from the passage of the Toxic Substances Control Act (15

U.S.C. §2601 et seq.) and the Resource Conservation and Recovery Act (42 U.S.C. §6901 et seq.) in 1976. And, of course, there was the tidal wave of reaction and litigation when the Comprehensive Environmental Response, Compensation, and Liability Act (42 U.S.C. §9601 et seq.) became law.

The vast majority of environmental articles, papers and educational courses have focused on the nature, extent, and ramifications of these "waves." No one, until now, has explored the "underlying current," the environmental reporting and recordkeeping requirements, present in all the major environmental statutes. Although many of the articles and courses have touched upon the subject in their treatment and analysis of a particular statute, no one (at least to our knowledge), has devoted an entire course to the study of environmental reporting and recordkeeping. This is unfortunate, since many of those in the regulated community have begun to realize the fact that if the "waves" do not get you, the underlying current probably will.

Perhaps one of the major reasons for this general lack of concern has been that until recently EPA has not placed reporting and recordkeeping violations high on its list of priorities. There had always been some limited enforcement of such requirements, but generally it was in conjunction with the enforcement of other provisions of a particular statute. For example, the failure to file annual reports under RCRA may have been included as a lesser charge to a violation for illegally disposing of hazardous wastes.

EPA's attitude with respect to reporting and recordkeeping violations is apparently changing. Perhaps as a result of the passage of Title III of the Superfund Amendments and Reauthorization Act of 1986 (*i.e.*, the Emergency Planning and Community Right-To-Know Act of 1986), which is in essence a reporting and notification statute, EPA is directing more and more of its resources and personnel to the sole enforcement of reporting and recordkeeping violations. Armed with a wide arsenal of civil and criminal sanctions, EPA has begun sending a message to the regulated community that the failure to keep appropriate records and to report information required under the each of the environmental statutes will result in enforcement actions -- administrative, judicial and criminal. Indeed, according to the EPA's Enforcement Accomplishments Report: FY 1989, enforcement activities in the area of reporting and recordkeeping are on a steady increase.

With the rise in EPA enforcement activity relating to reporting and recordkeeping violations, the regulated community has increased its activities to ensure compliance with the multitude of such regulations. The purpose of the course materials which follow is to present in one volume the major environmental reporting and recordkeeping requirements.

Aside from avoiding potential liability, fines, and criminal sanctions, companies may derive two additional major benefits by complying with such environmental reporting and recordkeeping requirements. First, compliance with such requirements can improve the environmental management effectiveness of a company. A comprehensive reporting and recordkeeping compliance program can identify weaknesses or gaps which may exist within the organization's informational network. By addressing those "blind spots," the company can develop a more effective and efficient approach towards managing its environmental activities, which in turn, would improve the company's overall compliance record and ability to respond in the event an environmental emergency should occur. A second major benefit of complying with environmental reporting and recordkeeping requirements is the increased level of comfort and security derived from knowing that the company has reduced its exposure to potential civil and criminal liability.

Our focus initially is on the specific reporting and recordkeeping requirements pertaining to each of the major environmental statutes and regulations promulgated thereunder. The materials that follow will examine these requirements.

THE CLEAN WATER ACT:
REPORTING AND RECORDKEEPING REQUIREMENTS

ROBERT L. COLLINGS
MORGAN, LEWIS & BOCKIUS

I. Introduction and Overview

The Clean Water Act (the Act)[1] authorizes and defines federal programs which regulate, among other things, direct discharges*[2] of pollutants* from point sources* to the surface waters* of the United States, indirect discharges* through publicly-owned treatment works* (POTWs) to surface waters, spills* of oil and designated hazardous substances* to surface waters, and activities which involve the dredging of materials within surface waters and the deposition of materials in such waters or in wetlands*. The programs mentioned here each contain requirements for reporting information concerning these activities or for preparing and maintaining records, and in some cases require filing records with the Environmental Protection Agency (EPA) or other federal agencies and their state "proxy" agencies.

---

[1] Codified at 33 U.S.C. § 1251 et seq.

[2] Words or phrases marked with an asterisk (*) are defined in the glossary attached as Appendix A.

All of these requirements are authorized by § 308 of the Act, 33 U.S.C. § 1318, which authorizes EPA to require the preparation and maintenance of records and the submission of reports as generally authorized under § 308, or as necessary to implement other sections of the Act, or as needed for rulemaking and enforcement activities.[3]

This chapter details the specific reporting and recordkeeping requirements of these programs, explains the regulations, forms and policies which apply, and discusses relevant penalties, sanctions and caselaw interpreting and enforcing the requirements.

II. Direct Dischargers - NPDES Requirements

The National Pollutant Discharge Elimination System (NPDES)[4] is a permit program which applies to anyone who adds any pollutant to surface waters from a point source. Generally speaking, all such persons must obtain permits from EPA or from a state agency which has received federal approval to conduct the NPDES program. Each NPDES permit issued to a facility may define one or more discrete point sources (outfalls) from which a discharge is authorized. The discharger's application must provide detailed information concerning the volume of each outfall's discharge and the constituents (pollutants) of the discharge. The information in the application is used to set effluent limitations in the permit for each outfall, based on one or more applicable standards. This information is also used to impose specific requirements for reporting and recordkeeping.

<u>NPDES Recordkeeping</u>. EPA regulations require each permit to contain certain recordkeeping requirements. These requirements are imposed in EPA permits and in all State-

---

[3] Many of these requirements may also be authorized by the general administrative rulemaking authority of § 501 of the Act, 33 U.S.C. § 1351.

[4] This program is authorized by § 402 of the Act, 33 U.S.C. § 1342.

issued NPDES permits. 40 CFR § 122.41(j)[5]/ specifically requires, as a condition in each permit, the maintenance of records of

(1) all monitoring information (see below)
(2) calibration and maintenance records and all original strip chart recordings for continuous monitoring instrumentation (such as flow, pH, or temperature, when required)
(3) copies of reports required by the permit (see below)
(4) records of all data used to complete the permit application.

Duration of Recordkeeping. Records must be kept at least 3 years from the date of the sample, measurement, report or application. 40 CFR § 122.41(j)(2). This recordkeeping period may be extended by request of the Director* at any time (no permit action is needed). The raw sample data should be kept until 3 years after the last submitted report or application referencing such data, as a matter of good recordkeeping; the legal obligation ends 3 years after data collection.

NPDES Monitoring Records. 40 CFR § 122.41(j)(3) requires records of the date, exact place and time of sampling or measurement, name of person taking sample, measuring or performing analyses, date(s) of analysis, techniques[6]/, and results. The Discharge Monitoring Report (DMR) discussed below does not require all of the information which must be kept in the discharger's records available for inspection or submission upon request. Monitoring requirements must specify proper installation, use

---

[5]/ References to sections of EPA rules, all of which are codified in Title 40 of the Code of Federal Regulations, are set forth as 40 CFR § ____.

[6]/ This includes both the sampling technique, e.g., grab sample or composite sample, manual or automated sampling, as well as the analytical technique, which is specified either in 40 CFR Part 136 or an alternate method approved by the permit issuer.

and maintenance, when appropriate, of monitoring equipment or methods (including biological monitoring methods when appropriate); type, interval and frequency of monitoring sufficient to yield representative discharge data including, when appropriate, continuous monitoring. See 40 CFR § 122.48. Monitoring requirements also include (see 40 CFR § 122.44(i)):

(1) mass or other specified measurement for each limited pollutant;

(2) volume of effluent discharged;

(3) <u>as appropriate</u>, measurements of internal waste streams under 40 CFR § 122.45(i); pollutants in intake water for net limits under 40 CFR § 122.45(f); frequency, rate of discharge, etc. for noncontinuous discharges under 40 CFR § 122.45(e), and measurements of "notification" pollutants not limited in the permit, see 40 CFR § 122.42(a) <u>and specific reporting, item (10), below</u>;

(4) Use of analytical methods specified in 40 CFR Part 136 or approved alternate methods;

(5) Reporting requirements (see below).

NPDES Reporting. General. A permittee may be required to submit, i.e., report, upon request by the NPDES Authority*, information needed for permit actions (modification, renewal or termination) or for a governmental determination of compliance. Copies of any of the records required above must also be submitted upon request. 40 CFR § 122.41(h).

Specific Reporting. 40 CFR §§ 122.41(l), (m) and (n), 122.42(a) and (b), 122.44(g), (i) and (j) and 122.48 all establish required permit conditions for reporting to the NPDES Authority, as follows:

(1) Planned physical changes to the permitted facility which may constitute a new source (subject to possibly more stringent new source treatment standards), or which may change the nature or quantity of pollutants. The latter requirement applies only to pollutants not

subject to effluent limits or to "notification levels" (see item 10 below). No specific reporting time is set, but the time of report must be reasonable in relationship to the timing of plans, presumably in advance of implementation.

(2) Advance notice of planned changes in the facility or activities either of which may cause noncompliance with permit requirements such as effluent limits or compliance schedules. No specific "lead time" is set. <u>See</u> item 1.

(3) Prior notice of transfers, which may trigger permit modifications or revocation and reissuance. No specific time prior to the transfer is required.

(4) Discharge Monitoring Reports - <u>see below</u>.

(5) Compliance Schedules. Most permits contain effluent limits which are immediately effective. If compliance with limits at a future date is allowed, permits usually contain schedules of actions required to achieve compliance. A written report of compliance or noncompliance is required within 14 days of each specified milestone date in the schedule. Permits with Individual Control Strategies (ICSs) to meet water quality standard-based requirements under § 304(l) of the Act will contain compliance schedules in many cases.

(6) 24 hour reports/5 day letters. Any permit noncompliance, usually a discharge of pollutants exceeding permit limits for authorized pollutants, must be reported orally within 24 hours with a followup written report within 5 days for the following:

(a) noncompliance which may endanger health or the environment.

(b) unanticipated bypasses* which exceed effluent limits. (<u>see Bypasses below</u>).

(c) upsets* which exceed effluent limits. (<u>see Upsets below</u>)

(d) exceedance of maximum daily discharge (effluent) limits for certain pollutants listed in permits. see 122.44(g). Most permits require reporting categorically for toxic pollutants*[7], hazardous substances[8], or pollutants identified as indicator pollutants[9] to control toxic pollutants or hazardous substances.

The NPDES authority may waive the 5 day followup written report after 24 hr. oral notice.

(7) All other instances of noncompliance with permit conditions other than information required in DMRs, or related to compliance schedules or 24 hr./5 day reports under (6), must be filed in writing with the DMRs.

(8) Any relevant facts which the permittee failed to submit in its permit application or a required report, or any incorrect information in an application or report, shall be submitted or corrected in a report to the permitting agency (NPDES Authority) promptly after the permittee becomes aware of its failure or error.

(9) Anticipated Bypasses must be reported at least 10 days in advance if possible. (See Bypasses below).

(10) Notification levels. Notice must be given as soon as possible for prior or anticipated occurrences which the permittees knows or has reason to believe would result in

(a) a routine or frequent discharge exceeding the following notification levels:
° for toxic pollutants not limited in the permit, 100 micrograms per liter (µg/l).

---

[7] Designated pursuant to § 307(a) of the Act. The list of toxic pollutants under this Act is codified at 40 CFR § 401.15.

[8] For purposes of this Act, the list is codified at 40 CFR Part 116. See definition of "hazardous substance" in the NPDES rules at 40 CFR § 122.2.

[9] For a discussion see NRDC v. EPA, 882 F.2d 104, 125 fn. 20 and related text (D.C. Cir. 1987).

- for acrolein or acrylonitrile <u>if not limited</u>, 200 µg/l.
- for 2,4-dinitrophenol or 2-methyl-4,6-dinitrophenol <u>if not limited</u>, 500 µg/l.
- for antimony <u>if not limited</u>, 1 milligram per liter (mg/l).
- for pollutants reported present at stated concentrations in the application <u>but not limited</u>, five (5) times the maximum concentration stated in the application.
- other notification levels set in the permit for pollutants <u>not limited</u>.

(b) <u>a nonroutine or infrequent discharge</u> of a toxic pollutant <u>not limited in the permit</u>, which will exceed the highest of the following:
- 500 micrograms per liter (µg/l)
- 1 milligram per liter (mg/l) of antimony
- ten (10) times the maximum concentration for the pollutant reported in the permit application, or
- other notification levels for the pollutant specified in the permit.

(11) (a) Operators of publicly owned treatment works (POTWs) must report any new introduction of pollutants by an indirect discharger if such pollutants would be subject to the Act's technology-based treatment standards when discharged directly.

(b) POTWs must also report any substantial change in the volume or character of pollutants introduced to the POTW by a source which is introducing pollutants at the time the POTW's NPDES permit is issued.

The information required under (a) and (b) is:
- The quantity and quality of effluent introduced (flow and pollutants), and
- any anticipated impact of the change caused by

(a) or (b) on the quantity or quality of effluent discharged from the POTW.

There is no specific timing for the notice, which must simply be "adequate".

(12) Results of required monitoring must be reported with a frequency dependent on the nature and effect of the discharge, in no case less than once per year.

(13) POTWs must report, either in permit applications or as required in (11) above, the character and volume of pollutants introduced by any significant indirect discharger subject to pretreatment standards promulgated under § 307(b) of the Act or subject to general pretreatment regulations at 40 CFR Part 403.

<u>Discharge Monitoring Reports (DMRs)</u>. In <u>NPDES Self-Monitoring System - User Guide</u> (EPA Office of Water Enforcement and Permits, March, 1985) EPA describes the DMR at page 5: "The DMR is a routine compliance report that gives a summary of the quality and/or quantity of the permittee's discharge. The DMR is submitted to the regulating agency in accordance with a schedule...in the permit, usually monthly or quarterly. It provides data on facility flow, sample collection, and analytical results. <u>It is extremely important that the data reported on the DMR be accurate and timely because the reported data will be compared with permit effluent limitations to determine facility compliance.</u>" (underlining added.)

DMR data is reported as a mass or concentration of pollutants. The highest sample value is generally reported as the daily maximum. While the arithmetic average of all samples analyzed in a given month is generally reported as the monthly average (a geometric average is used for bacterial pollutant parameters).

It is important to note that analytical results of all samples taken at the specified monitoring points must be incorporated in the DMR if the analysis was done using the required methods. The specified minimum number of samples must be collected at the proper times and locations by the right methods and analyzed using the approved methods. Inclusion of additional data depends on whether the methods of sample collection and analysis render the data compatible with normal permit data and whether the discharge sample is representative of the normal discharge during plant operations. Concerns about inclusion of specific added data should be addressed to the NPDES authority in time to resolve them prior to the reporting deadlines.

DMR data is the principal tool used by EPA, by the States, and by private citizens or citizen groups, to enforce compliance with the Act. The discharge numbers reported on a DMR constitute at best a presumption of violation and liability when they exceed the effluent limits in the permit (which are printed on the DMR for convenience); at worst, the discharger is allowed no opportunity to challenge the data in an enforcement proceeding. The failure to submit a complete DMR is a separate violation. (See Enforcement, below).

A sample DMR form with instructions is attached as Appendix B.

Bypasses. A bypass is the intentional direct discharge of pollutants without full treatment. It includes discharges of untreated or partially treated wastewater. There are two classes of bypass:

(1) bypasses which do not result in discharges exceeding effluent limits are allowed when the bypass is for essential maintenance to assure efficient treatment operations. No notice to, or approval of, the NPDES authority is required.

(2) bypasses, either anticipated or unanticipated, which exceed effluent limits are prohibited unless the permittee meets 3 criteria, and, in the case of an anticipated bypass, obtains advance approval from the NPDES authority. The criteria are:
  (a) Notice/Reporting
    • For known, anticipated bypasses, at least 10

  days prior to bypass (noted in <u>Specific Reporting</u>, item 9).

- For unanticipated bypasses, 24 hour oral notice (noted in <u>Specific Reporting</u>, item 6(b).

(b) The bypass was unavoidable to prevent loss of life, personal injury or "severe property damage" (defined at 40 CFR § 122.41(m)(1)(ii)).

(c) There were no feasible alternatives such as auxiliary treatment, retention of wastes, maintenance during normal downtimes. If the bypass occurs during normal equipment downtime or preventive maintenance, the permittee must show that "reasonable engineering judgment" would not have dictated the prior installation of back up equipment to prevent exceedance of limits.

These bypass rules have been fully reviewed and approved by the courts. <u>NRDC v. EPA</u>, 822 F.2d 104, 122-126 (D.C. Cir. 1987).

<u>Upsets</u>. These are "exceptional" incidents in which "unintentional and temporary noncompliance with technology-based permit effluent limitations" occurs for reasons "beyond the reasonable control of the permittee". 40 CFR § 122.41(n)

(1) <u>Effect</u>. If the permit contains an upset provision, a properly demonstrated upset is a defense to enforcement actions.

- <u>Note</u> that States are free to deny this defense by omitting the permit condition. It is not required in State programs (see 40 CFR § 123.25, NOTE) which may be more stringent.

(2) <u>Exclusions</u>. The defense applies only to enforcement of technology-based limits. Upsets are not a defense to noncompliance with water quality based effluent limits. This exclusion was upheld in <u>NRDC v. EPA</u>, 859 F.2d 156, 205-210 (D.C. Cir. 1988). Also excluded are claims of upset based on operator error, improperly designed or

inadequate treatment, lack of **preventive** maintenance or careless operation.

(3) <u>Recordkeeping</u>. The permittee bears the burden of proof and must have properly signed, contemporaneous operating logs or other records or evidence proving an upset which caused the noncompliance, during proper operation, with required notice and proper remedial responses taken.

(4) <u>Notice/Reporting</u> - 24 hour notice. <u>See Specific Reporting</u>, item 6(c) <u>above</u>.

<u>NPDES Noncompliance Reports (NCRs) - Contents</u>. Both the 5 day noncompliance letters described in <u>Specific Reporting</u>, item 6 (unless waived after 24 hour oral notice), and written reports of other noncompliance filed with the DMRs (<u>Specific Reporting</u>, item 7) must contain:

(1) description of noncompliance and cause;

(2) period of noncompliance (exact times);

(3) anticipated time of further noncompliance, if not yet corrected, and

(4) steps taken or planned to reduce, eliminate and prevent reoccurrence of the noncompliance.

NCRs are another primary enforcement tool under the NPDES program. As required government reports which must be certified as correct, the provision of false or misleading information is punishable by law. Certification and signing requirements are specified at 40 CFR §122.22, and are required permit conditions under 40 CFR § 122.41(k). See <u>Enforcement</u>, <u>below</u>.

III. Indirect Dischargers and POTWs - Pretreatment Requirements

EPA's general pretreatment regulations[10] govern many persons who discharge wastewater into POTWs[11] for treatment and discharge to surface waters. These rules contain federal-

---

10/ 40 CFR Part 403.

11/ Discharges include putting pollutants into sewers connected to POTWs, or transporting pollutants by truck, rail, pipeline or other means to a POTW.

-15-

reporting requirements for indirect dischargers (40 CFR § 403.12), and specific requirements for State or local (POTW) pretreatment programs imposing record-keeping and reporting duties on indirect dischargers. Other rules require reports by the POTW. The State or POTW may use its authority under State or local laws to develop additional recordkeeping or reporting requirements necessary for its compliance with federal, state or local laws. Such individual state and local requirements may vary, and are not reviewed here.

Exclusions. The federal requirements do not apply to domestic wastewater or to wastewater discharged to sewers not connected to POTWs. The latter are considered direct discharges and the discharger is subject to NPDES requirements. (Sewers owned and operated by, and connected to POTWs may have overflow points prior to treatment. Indirect dischargers authorized by the POTW to use such sewers are not considered direct dischargers. POTWs are responsible for such discharges at this time).

Federal Reporting Requirements

(1) Baseline Reports. Within 180 days after a federal pretreatment standard is promulgated for an industrial category, or 180 days after a final EPA response to request for determination that a standard applies, when the final response applies the standard, the covered indirect discharger must file a baseline report if it currently discharges or is scheduled to discharge. New indirect dischargers must file a baseline report 90 days prior to commencement of discharge. The reports are filed with the Control Authority*. The baseline report must contain (applicability indicated in parentheses):

  (a) -name and address of facility, operators and owners (existing and new sources);
  (b) a list of all facility environmental permits (existing and new sources);
  (c) a brief description of the nature and average rate

of production processes, SIC codes*, and a schematic showing points of discharge to POTW. (existing and new sources);

(d) information showing measured daily average and maximum daily flows to the POTW (or verifiable estimates, if allowed by the Control Authority based on cost or feasibility factors) from each wastewater stream covered by a standard, and any other wastewater stream containing pollutants of the type regulated by the standard (and therefore requiring use of EPA's combined wastestream formula to determine compliance - see 40 CFR § 403.6(e)). (existing sources; new sources must report estimates);

(e) identification of pretreatment standards applicable to each process. (existing and new sources);

(f) analytical results of representative discharge samples showing the nature and concentration (or mass, when required) of regulated pollutants from each regulated process as a daily maximum and daily average. (existing sources; estimates by new sources);

- at least 4 grab samples are required for pH, cyanide, total phenols, oil and grease, sulfide and volatile organics where regulated. Flow-proportionate 24 hour composite samples are required for all other regulated pollutants unless waived by the Control Authority as infeasible, in which case 4 grab samples are required or more as needed to provide a representative sample. At least one sample is required in all cases. Sample locations must be downstream of any pretreatment or from the regulated process if there is no pretreatment. Other sampling must be done as required to use the combined wastestream formula to show compliance. (existing sources; estimates by new sources);

(g) Approved alternate limits must be identified (existing and new sources);

(h) the time, date and place of sampling, methods of analysis, and certification that samples are representative of normal work cycles and expected discharges. (existing sources);

(i) statement by an authorized representative*, which is certified by a qualified professional, that standards are being consistently met, or a certified statement of additional pretreatment and/or operation and maintenance needed for consistent compliance. (existing sources);

(j) The shortest practicable compliance schedule, if necessary, to meet standards on or before the required compliance deadline. See compliance schedule requirements at 40 CFR § 403.12(c).

• Where standards or limits are modified, an amended statement of compliance or compliance schedule is due within 60 days. (existing sources).

Historical data, as opposed to newly generated samples, may be used for (4) or (6) when approved by the Control Authority. All analyses must use methods in 40 CFR Part 136 or approved alternate methods.

(2) Compliance Schedule Reports. Not later than 14 days after each compliance schedule milestone and after the compliance deadline, a progress report is due, stating

(a) whether the milestone or deadline was met, and

(b) if not, the expected date of achievement, reason for delay, and steps to restore compliance to the schedule.

Reports are due each 9 months at a minimum.

(3) Compliance with Standards. Must be reported within 90

days after the deadline (or after commencement of a new source discharge).

The report must contain current measurements of information as specified in the baseline report items (d)-(i), above. If the standards are based on equivalent mass (measured concentration times long-term flow), a reasonable measure of long term production rates shall be included for comparison with production rates reported in the baseline report. Otherwise, actual production during sampling shall be reported for comparison with the baseline report.

(4) Periodic Compliance Reports. After the compliance date, the indirect discharger must report to the Control Authority in June and December of each year, or on different or more frequent dates specified by the Control Authority or Approval Authority, providing the following:

  (a) nature and concentration of each regulated pollutant, and measured or estimated daily average and daily maximum flows for the reported period, for each discharge subject to standards. More detailed flow reporting may also be required.

  (b) mass of regulated pollutants discharged, where standards are expressed as mass.

  (c) where equivalent mass limits are based on long-term flows, long-term production rates; otherwise, actual production.

  (d) sampling and analysis must occur during the reporting period and must be representative. More frequent monitoring data using approved methods must be included.

(5) Potential Problems. Indirect dischargers must immediately report to the POTW any dischargers which may cause damage to the POTW or interfere with its operations and compliance, including "slug loads". This requirement applies to _all_ dischargers, whether or not subject to a federal categorical pretreatment standard.*

(6) <u>Exception</u>. The POTW may itself perform all sampling required for items (1), (3) and (4), in which case those reports are not required.

(7) 24 hour Reports. The indirect discharger must report to the Control Authority within 24 hours after becoming aware of a violation, e.g., receiving analytical reports or information indicating a violation.

(8) 30 day resampling after 24 hour report. Resampling and reporting of results must occur within 30 days unless the Control Authority performs sampling at least once per month, or the Control Authority samples between the time that the discharger initially sampled and the time analytical results are received.

(9) Indirect dischargers not subject to categorical Pretreatment Standards must be required to provide appropriate reports. The requirements are not specified. Presumably, they must be reasonably needed for the POTW to control such dischargers, so as to avoid interference, damage or pass through resulting in noncompliance by the POTW direct discharge.

(10) Each discharger must notify the POTW in advance of any substantial change (not defined) in the volume or character of pollutants discharged to the POTW. This is tied to the POTW's NPDES duty. See NPDES Specific Reporting, item 11(b), above.

All sampling and analysis required above must be informed using methods in 40 CFR Part 136 or EPA-approved methods. Baseline and compliance reports must be signed by a specified corporate officer, general partner, proprietor or duly authorized representative who certifies, under penalty of law, the accuracy of reported information. 40 CFR § 403.12(1).

<u>Indirect Discharger Recordkeeping</u>. 40 CFR § 403.12(o). Records must be kept of the date, exact time and place of sampling, persons performing sampling, measurements or analyses, dates and methods of analyses and results. This is

the same as the direct discharger rule at 40 CFR § 122.41(j)(3) - see NPDES Monitoring, above. Records must be kept of all monitoring activities (whether or not required) for 3 years, or longer if litigation exists involving the discharger or the POTW, or if requested by the state or EPA. All records must be available for inspection and copying.

IV. POTW Requirements - Pretreatment

In addition to reports or records required by the POTW's NPDES permit, the POTW is subject to added reporting and recordkeeping pretreatment rules.

Pretreatment Reports. POTWs with approved pretreatment programs must file annual reports describing their program activities, beginning in the first year after approval. POTWs are not subject to this report if no Pretreatment Program is required, i.e., if the total design flow is ≤ 5 million gallons per day (mgd) or if the POTW does not receive industrial indirect discharges which may interfere with or pass through the POTW or which are subject to categorical pretreatment standards. The EPA or a State Approval Authority may require a Pretreatment Program for POTWs excluded above if they decide that interference or pass through may occur based on the nature or volume of industrial influent, treatment upsets, POTW permit violations, contaminated POTW sludge or other circumstances. The reports must contain:

(1) an updated list of indirect dischargers identified as subject to categorical pretreatment standards or local limits or both. Applicable standards or limits for each discharger must be specified, and list deletions explained.

(2) a summary of discharger compliance status for the period.

(3) a summary of POTW pretreatment compliance and enforcement activities, including inspections, during the period.

(4) any other relevant information requested by the Approval Authority.

Report of compliance with Pretreatment Program Development Schedule. POTWs must report within 14 days after each compliance schedule milestone date on progress in meeting the schedule, including:

(a) Whether or not the schedule date was met, and
(b) if not, date of expected compliance, reasons for delay, and steps to resume the schedule.

POTW recordkeeping. In addition to NPDES recordkeeping requirements, POTWs must keep all baseline and compliance reports of each indirect discharger for 3 years, or longer if enforcement litigation exists, or as requested by EPA or an Approved State.

Signatory requirements. POTW pretreatment annual reports must be signed by a principal executive officer, ranking elected official or a duly authorized employee responsible for overall operation of the POTW.

V. Clean Water Act Spill Reporting and Spill Prevention Reporting and Recordkeeping.

Spill Reporting and Prevention. Under § 311(b)(5) of the Act, any owner or operator of a vessel, onshore or offshore facility must report any discharge of oil or hazardous substance in harmful quantities: into the navigable waters of the United States; onto adjoining shorelines; into the contiguous zone[12]; which may affect natural resources owned or managed by the U.S. under the Magnuson Fishery Conservation and Management Act of 1976[13], or discharged in connection with activities subject to the Outer Continental Shelf Lands Act or Deepwater Port of 1974.

---

[12] unless a vessel discharge is authorized by Marpol 73/78, the International Convention for the Prevention of Pollution from Ships, 1973, as modified by the Protocol of 1978 relating thereto, annex I regulatory oil discharges (effective 10/2/83).

[13] id.

Basically, any discharge of a potentially harmful quantity into surface waters or adjoining lands subject to U.S. jurisdiction must be reported.

(1) Designation of Hazardous Substances. Hazardous Substances are listed at 40 CFR Part 116.

(2) Harmful Quantities.

    (a) <u>Oil</u>. Any quantity which violates an applicable water quality standard or causes an iridescent sheen on the surface is potentially harmful. 40 CFR §§ 110.3-110.5 and 110.11.

    (b) <u>Hazardous Substances</u>. Discharge of a reportable quantity, as listed at 40 CFR § 117.3, is deemed potentially harmful.

(3) <u>Exceptions</u>

    (a) Hazardous Substance Reporting. Discharges subject to and complying with various regulatory programs or permits including NPDES permits, RCRA, FIFRA (the federal pesticide law) and the Marine Protection Research and Sanctuaries Act and discharges in compliance with orders of an EPA or Coast Guard On-Scene Coordinator. See 49 CFR §§ 117.11 and 117.12.

    (b) Oil. Discharges subject to and complying with NPDES program requirements and permits, see 40 CFR § 117.12, discharges of oil contained in dredged spoil, and discharges authorized under the Deepwater Ports Act or Marpol 73/78 Annex I.

(4) Discharges to POTWs (hazardous substances only). Discharges of reportable quantities of hazardous substances from mobile sources to POTWs must be reported unless the POTW gives prior approval and the discharge has been treated to meet applicable pretreatment standards and to comply with discharge limits of the POTW.

(5) Reporting Procedures. EPA and the Coast Guard share jurisdiction over reporting on a roughly inland/coastal

waters, lakes and major rivers split. EPA rules at 40 CFR § 110.10 (oil), and § 117.21 (hazardous substances), and Coast Guard at 33 CFR § 153.203 all require immediate reporting:

(a) To the National Response Center (NRC) hotline at 1-800-424-8802 (D.C. area 202-267-2675(426?))

(b) If (a) is not practicable, to the EPA or Coast Guard On-Scene Coordinators for the Geographic Area (see 33 CFR Part 153, subpart B, Table 1), or

(c) If (a) and (b) are not practicable to the nearest Coast Guard Unit.

(d) In cases (b) and (c), a report to the NRC must be make by the owner or operator as soon as possible thereafter.

EPA rules also reference procedures under the National Oil and Hazardous Substances Pollution Contingency Plan (NCP). These procedures at 40 CFR § 300.51(b) (oil) and § 300.63(b) (hazardous substances) are identical to the above procedures.

(6) Contents of Reports. While not specified in the rules cited above, the NRC ordinarily requests the same information required under CERCLA § 103(a) or SARA Title III § 304 reports, as follows:
   (a) name and chemical identity of substance;
   (b) estimate of quantity;
   (c) time and duration of release;
   (d) location of release;
   (e) known or anticipated acute or chronic health risks;
   (f) medical advice for exposures, if appropriate;
   (g) precautions taken or responses;
   (h) names and telephone numbers of persons reporting and contacts for further information.

(7) <u>Recordkeeping</u>. There are no recordkeeping requirements under these rules.

Spill Prevention Control and Countermeasure Plans (SPCC).

EPA has issued rules under § 311(j) of the Act and its general rulemaking authority under § 501 of the Act, which require certain onshore and offshore facilities to prepare SPCC plans to prevent and respond to oil spills.

(1) Reporting. The rules at 40 CFR Part 112 have no specific reporting requirements in addition to oil spill reporting requirements described above. Plans may contain additional reporting requirements relating to seeking local assistance or emergency reporting and response.

(2) Recordkeeping. The SPCC Plan itself, when required, must be maintained as a record at all times available for reference and inspection. The plan requires recordkeeping of all required facility inspections, of releases of stormwater from certain containment areas, and of oil production well shut-in valves and devices.

(3) Exclusions. Vessels and transportation-related facilities subject to Coast Guard jurisdiction, facilities which cannot reasonably discharge (ignoring man-made preventive features) to surface waters or shoreline, and facilities with oil storage only in underground storage $\leq$ 42,000 gallons and aboveground storage $\leq$ 1,320 gallons total and each aboveground storage container $\leq$ 660 gallons.

Oil Pollution Prevention Regulations for Transfer Facilities, Public and Private Vessels.

The U.S. Coast Guard rules at 33 CFR Parts 154, 155 and 156 require reporting of oil discharges as provided in oil transfer procedures (§ 155.750(a)(9)), and advance reporting 4 hours prior to oil transfer for facilities, as required by the Captain of the Port (COTP), which are mobile, in remote locations, have prior spill histories, or conduct infrequent transfers. 33 CFR § 156.118(a). Lightering (vessel to vessel transfers) may also require 4 hours advance notice. 33 CFR § 156.118(b).

Facilities transferring more than 250 barrels per operation must keep as available records the operations manual and also:
 (1) the letter of intent to transfer
 (2) names of persons designated in charge
 (3) date and result of required equipment inspections
 (4) hose information and
 (5) COTP inspections for 3 years. 33 CFR § 154.740.

Public Vessels receiving more than 250 barrels of oil may be required to keep as records the Declaration of Inspections, names of persons in charge, dates and results of equipment inspections, and hose information. 33 CFR § 155.820.

VI. Dredged Materials Discharges and Filling Waters and Wetlands.

Under § 404 of the Act, the U.S. Army Corps of Engineers administers a permit program regulating discharges of dredged materials or fill materials into waters of the United States or wetlands. EPA issues guidance for certain aspects of these programs and may review and veto certain permits. Generally, the permit program requires no special reports[14] or recordkeeping, other than requirements in Corps' permits to report any historical or archaeological remains not identified in the permitting process.

VII. Enforcement.

Violations of recordkeeping and reporting requirements are enforceable under § 309 of the Act by administrative compliance orders, civil or criminal suits, and judicial or administrative penalties. As an alternative sanction, §§ 402 and 404 of the Act authorize permit revocation for violation of conditions. § 311(b)(5) of the Act authorizes criminal fines for failure to report spills, and § 311(j)(2) authorizes civil penalty assessments for violation of EPA and Coast Guard spill prevention requirements. The amounts of

---

[14] Application materials are not considered reports.

the penalties and procedures for assessment are as follows:
(1) NPDES & Pretreatment violations - administrative penalties
   (a) Class I: $10,000 per violation (each day is a separate violation), up to $25,000 per penalty order. Notice of proposed action must be given, and an appeal using non-APA[15]/ procedures with reasonable opportunities to be heard and present evidence. Judicial review (§ 309(g)(8)) may be sought on the hearing record applying a substantial evidence/abuse of discretion test.
   (b) Class II: $10,000 per day of violation up to $125,000. Notice must be given, and a formal APA hearing is allowed. See 40 CFR Part 22 procedures. Judicial review is the same as (a).
   (c) Also applies to failure to report historical sites under § 404 permits.
(2) NPDES & Pretreatment violations - civil penalties.
   (a) up to $25,000 per day for each violation.
   (b) continuing violations on multiple days are assessable. Chesapeake Bay Foundation, Inc. v. Gwaltney of Smithfield, Ltd., 791 F.2d 304, 314-315 (4th Cir. 1986) vacated on other grounds, __ US __, 108 S.Ct. 326, 98 L.Ed. 2d 306 (1987).[16]/
   (c) EPA's Civil Penalty Policy may be applied to set penalty based on economic benefit of violation, seriousness of violation, compliance record, economic impact and harm. Gwaltney, 611 F.Supp. 1542 (E.D.Va. 1985).
(3) Negligent Violations - NPDES/Pretreatment (criminal)
   (a) First offense - $2500 to 25,000 per day of violation and/or imprisonment up to 1 year.

---

[15]/ Hearing provisions of the Administrative Procedures Act, 5 USC §§ 554 or 556.

[16]/ See also, opinion on appeal after remand, 890 F.2d 690, 698 n.7.

- (b) Second offense after a conviction - up to $50,000 per day of violation and/or imprisonment up to 2 years.
- (4) Knowing violations - NPDES/Pretreatment (criminal).
    - (a) First offense - $5,000 to $50,000 per day of violation and/or imprisonment up to 3 years.
    - (b) Second offense after conviction - up to $100,000 per day of violation and/or imprisonment up to 6 years.
- (5) False statements in reports or records - NPDES/Pretreatment.
    - (a) Applies to knowing and material false statements and to falsification of, or tampering with, or rendering inaccurate any monitoring device or method related to such reports or records.
    - (b) First offense - fines up to $10,000 and/or imprisonment up to 2 years.
    - (c) Second offense after conviction - up to $20,000 per day of violation and/or up to 4 years imprisonment.
- (6) Spill Reporting Violations - § 311(b)(5) (criminal)
    - (a) fines up to $10,000 and/or imprisonment up to 1 year.
    - (b) May be enforced by EPA or the Coast Guard through court proceedings.
    - (c) Actual notification may not be used in any criminal proceeding, e.g., to show late reporting.
- (7) Spill Prevention Regulations - § 311(j)(2).
    - (a) Civil (judicial) penalty up to $5,000.
    - (b) EPA compromises such penalties through administrative settlements.
- (8) § 404 violations - there are no instances known to the author of enforcement for failure to report historical remains. Presumably such violations would be subject to the § 309 enforcement procedures described in (1) - (4) above.

Discussion of Enforcement

The courts have generally held that § 308 of the Act and the NPDES rules require direct dischargers to monitor their discharge and provide a complete and accurate record. Sierra Club v. Simkins Industries, Inc, 847 F.2d 1109, 1111-12 (4th Cir. 1988) cert. den. -- US--, 109 S.Ct. 3185, 102 L.Ed.2d 580 (1989); U.S. v. B.P. Oil, 1989 WL 83623 (E.D.Pa. 1989). Simkins upheld a penalty of $977,000 ($1000 per violation) for 977 days of failure to sample, analyze, keep records or submit and maintain DMRs. See also, Menzel v. County Utilities Corp., 712 F.2d 91, 94-5 (failure to file reports violates the Act).

DMRs and noncompliance reports may be used to conclusively establish exceedance of permit limits. "In short, a discharger must report its own permit violations should they occur." SPIRG v. Fritzche, Dodge & Olcott, 579 F.Supp. 1528, 1531 (D.N.J. 1984), affirmed, 759 F.2d 1131 (3d Cir. 1985). They "may be used as admissions to establish...liability", SPIRG v. Monsanto, 600 F.Supp. 1479, 1485 (D.N.J. 1985); Sierra Club v. Union Oil Co. of Cal., 813 F.2d 1480, 1492 (9th Cir. 1987); U.S. v. M.D.C., 23 Env. Rep. Cas. (BNA) 1350, 1355, 16 Envt'l L. Rep. 20,621, 1985 WL 9071 (D.Mass. 1985). A few cases have allowed defenses to be asserted. U.S. v. City of Moore, 1985 WL 6115 (W.D.Ok. 1985) (defendant may prove inaccurate testing by outside laboratory showing false violations)[17]/; Sierra Club v. Kerr-McGee, 23 Env. Rep, Cas (BNA) 1685, 1985 WL 6029 (W.D.La. 1985) (summary judgment denied based on upset defense); U.S. v. Shell Oil Co., 817 F.2d 1169, 1174 (5th Cir. 1987) (bypass defenses); Chesapeake Bay Foundation v. Bethlehem Steel Corp., 608 F.Supp. 440, 453 (D.Md. 1985) (57 violations are possible upsets - no summary judgment); Friends of the Earth v. Facet Enterprises, Inc., 618 F.Supp. 532, 532-6 (W.D.NY. 1984) (allowing upset defense). Generally, the inaccuracy of

---

17/ The Contractor defense was rejected in SPIRG v. Monsanto, supra.

testing or scientific error has not been allowed as a defense. SPIRG v. Fritzche, Dodge & Olcott, supra at 1538-9; Atlantic States Legal Foundation v. Al Tech Specialty Steel Corp., 635 F.Supp. 284, 289 (N.D.NY 1986); NRDC v. Outboard Marine Corp., 702 F.Supp. 690 (N.D. Ill. 12/28/88) (permittee must challenge inaccurate test method by appeal rather than reporting with a qualification as to accuracy). Upset defenses to suits based on DMRs require an affirmative showing that the permit conditions allow the defense, U.S. v. B.P. Oil, supra (no permit upset condition); Connecticut Fund for the Environment v. Upjohn, 660 F.Supp. 1397, 1416 (D.Ct. 1987) (no proof of defense); SPIRG v. AT&T Bell Laboratories, 617 F.Supp. 1190, 1204 (D.N.J. 1985) (no permit condition for upsets). SPIRG v. Jersey Central Power and Light, 642 F.Supp. 103, 108-9 (D.N.J. 1986) (not temporary problem, thus no upset); SPIRG v. Georgia Pacific, 615 F.Supp. 1419, 1431 (D.N.J. 1985) (too many violations for upset).

The failure to file reports has also been used not as a violation, but rather as a basis for tolling the statute of limitations. Atlantic States v. Al Tech, supra at 289, on the theory that you must be able to see the required reports in order to exercise citizen enforcement rights. Accord FOE v. Archer Daniels Midland, Co., 1986 WL 13541 (N.D.NY 1986).

Criminal liability has been imposed on corporations for filing false reports, and on the involved employees. U.S. v. Olin Corp., 465 F.Supp. 1120 (D.NY. 1979); U.S. v. Little Rock Sewer Committee, 460 F.Supp. 6 (W.D.La. 1978). Penalties may be assessed under § 309(c) of the Act for falsification of any required report, and under 18 U.S.C. § 1001 for any other false report filed with the government (whether or not required).

In the area of § 311(b)(5) spill reporting, the statutory duty to report and penalty was upheld first in U.S. v. Boyd, 491 F.2d 1163 (9th Cir. 1973). A few cases have been reported which addressed the required timing for immediate reporting. U.S. v. Kennecott Copper Corp., 523

F.2d 821 (9th Cir. 1975) (spill of 173,000 gallons on 11/30 reported 12/3 - violation); U.S. v. Ashland Oil Co., 504 F.2d 1317 (6th Cir. 1974) (spill known at 7 p.m., report next day at 10:10 a.m. - violation); U.S. v. Messer Oil, 391 F.Supp. 557 (W.D.Pa. 1975) (spill in remote area at small facility over Memorial Day weekend, known on 5/27 but not reported by 5/31 may be a violation, but dismissed due to technical error in information). Corporations can be held liable for violations by their employees. Apex Oil v. U.S., 530 F.2d 1291 (8th Cir. 1976); U.S. v. Hougland Barge Line, Inc., 387 F.Supp. 1110 (E.D.Pa. 1974). Even low ranking employees may be individually liable if they are the "person in charge" of a discharging facility. U.S. v. Carr, 880 F.2d 1550 (2d Cir. 1989).

The reports of oil or substance spills may not be used in any criminal proceedings but may be used to assess civil fines for a prohibited spill. U.S. v. Ward, 448 U.S. 242, 100 S.Ct. 2636, 65 L.Ed.2d 742 (1980). However, one court had earlier held that a late report is not immediate notice as required by statute, so there is no protection from use of the report in criminal proceedings. U.S. v. Ashland Oil, 364 F.Supp. 349 (W.D.Ky. 1973). I disagree with this approach, which discourages late reporting and may harm the environment.

Conclusion

The failure to maintain required records and file reports has not been given much attention by EPA, except as an addendum to effluent violation cases. Likewise, Simkins stands as a single reported decision on a "filing" violation. On the other hand, a deliberate refusal to file is potentially subject to criminal sanctions, which is intended to outweigh any putative benefit from keeping your public file clear of effluent violation records. Many of the requirements described above are interconnected, and do not involve repetitive sampling or field investigations. One of the more troubling issues is the refusal of the courts to allow actual scientific disputes to be raised, as was done in Outboard Marine, supra. Requiring dischargers to raise every technical issue in a permit appeal creates an incredible burden. Many minor disputes need not be resolved in advance with the government, if you can "fine tune" the permit requirements when problems arise. A much better solution would be to allow disputes as to scientific accuracy to be preserved by comment on the record during permit issuance, and only require active appeal when the degree of error amounts to a potential for a significant "false positive".

The increasing use of unimpeachable administrative records and reports for enforcement as well as compliance purposes must inevitably lead to organized activity to monitor EPA's development of test methods as well as discharge standards. Some collective review of EPA's activities in that area would be useful as a protection against unreasonable enforcement actions in the future.

APPENDIX A

Glossary/References

Authorized Representative: the person who is approved, using proper company and government procedures, for the management and control of any regulated facility or activity, including the signing of required reports and the collection and custody of records.

Bypass: a planned or unplanned decision to discharge wastewater without all required treatment normally used. Effluent limits may or may not be exceeded during bypass. Under certain conditions a bypass may not be a violation even if limits are exceeded.

Categorical Pretreatment Standard: a treatment standard applicable to indirect dischargers of wastewater from specified industrial activities, resulting in the treatment of wastewater prior to treatment and discharge by a POTW. Intended to protect the POTW from damage to its treatment, to prevent contamination of municipal sludge, and to assure equivalent effluent quality from direct and indirect dischargers of the same wastewater.

Control Authority: A POTW or local body which has been approved under the Clean Water Act to manage its own pretreatment activities and indirect dischargers jointly with the Approval Authority. If neither the POTW nor any local body is approved, the Control Authority is EPA or the State, if a State agency has been approved by EPA to implement and manage the pretreatment program.

Direct Discharge: the addition of any pollutant to surface waters of the United States from a point source.

Director: the person who is in charge of and responsible for all actions of a state agency which has been approved by EPA to operate the NPDES program for that state.

Hazardous Substance: a substance designated in accordance with the procedures in section 311(b)(2)(A) of the Clean Water Act for purposes of that section. Also defined in section 309(c)(7) for purposes of criminally penalizing discharges to sewers.

Indirect Discharge: the introduction of any pollutant to a publicly owned treatment works for treatment and direct discharge, whether such introduction occurs through sewers, pipelines, or delivery by truck, rail or other means.

NPDES Authority: EPA, or a state agency approved by EPA in accordance with § 402(b) of the Clean Water Act, which operates the permit issuance, compliance/supervision and enforcement aspects of the NPDES program.

Point Source: any discernible, confined and discrete conveyance, including but not limited to any pipe, ditch, channel, tunnel, conduit, well, discrete fissure, container, rolling stock, concentrated animal feeding operation, vessel or floating craft, excluding agricultural stormwater, agricultural drainage and irrigation return flows.

Pollutant: defined in the Act as dredged spoil, solid waste, incinerator residue, sewage, garbage, sewage sludge, munitions, chemical wastes, biological materials, radioactive materials, heat, wrecked or discarded equipment, rock, sand, cellar dirt and industrial, municipal and agricultural waste, but excluding sewage from vessels, material injected into wells for production or disposal purposes approved by the State. Generally, anything discharged into surface waters other than water.

Publicly Owned Treatment Works: a treatment and (usually also a) direct discharge facility designed, constructed and operated using public funds and receiving primarily the sewage and wastewater of citizens and businesses within the jurisdictional area of government service provided by its elected or appointed managing body.

SIC Codes: the numerical codes published by the U.S. Department of Commerce and used to define major industrial activities of a place of business by group.

Spill: this term is used in this article to distinguish activities regulated under § 311-oil and hazardous substance discharges, from the discharges regulated by the NPDES program (direct discharges) and Pretreatment Program (indirect discharges). As used in this article, I mean "spilling, pumping, leaking, pouring, emitting, emptying or dumping which is not excluded as an authorized NPDES direct discharge or an indirect discharge excluded by § 311 rules.

Surface Water: a shorthand method of distinguishing water bodies and discharges, spills, or actions subject to the Clean Water Act. Generally, most water bodies with a surface which is visible from the ground surface are surface waters, and underground waters are not. Applies to discharges of water transported underground by sewers, and includes materials discharged to surface waters after travelling underground in a point source. Puddles and unchanneled, transient flows are not considered surface waters.

Toxic Pollutant: any pollutant listed by EPA as toxic in accordance with the procedures for evaluation specified in section 307(a) of the Clean Water Act, 33 USC § 1317(a).

Upset: an exceptional, temporary event beyond the control of a discharger which results in an exceedance of effluent limitations. It is an affirmative defense to liability when a permit condition so provides, and when the specified conditions for the defense are met.

Wetlands: any area which is periodically inundated so that its soils possess hydric characteristics or it supports hydrophilic vegetation. More precise definitions and requirements apply under federal and state programs.

NATIONAL POLLUTANT DISCHARGE ELIMINATION SYSTEM (NPDES)
DISCHARGE MONITORING REPORT (DMR)

PERMITTEE NAME/A&S (include
Facility Name/Location if different)

NAME _____

ADDRESS _____

FACILITY _____

LOCATION _____

Form Approved
OMB No. 2040-0004
Approval expires 6-30-88

PERMIT NUMBER (2-16) | DISCHARGE NUMBER (17-19)

MONITORING PERIOD

| FROM | YEAR (20-21) | MO (22-23) | DAY (24-25) | TO | YEAR (26-27) | MO (28-29) | DAY (30-31) |

NOTE: Read instructions before completing this form.

PARAMETER (32-37)		QUANTITY OR LOADING (3 Card Only) (46-53) / (54-61)		UNITS	QUALITY OR CONCENTRATION (4 Card Only) (38-45) / (46-53) / (54-61)			UNITS	NO. EX (62-63)	FREQUENCY OF ANALYSIS (64-68)	SAMPLE TYPE (69-70)
		AVERAGE	MAXIMUM		MINIMUM	AVERAGE	MAXIMUM				
	SAMPLE MEASUREMENT										
	PERMIT REQUIREMENT										
	SAMPLE MEASUREMENT										
	PERMIT REQUIREMENT										
	SAMPLE MEASUREMENT										
	PERMIT REQUIREMENT										
	SAMPLE MEASUREMENT										
	PERMIT REQUIREMENT										
	SAMPLE MEASUREMENT										
	PERMIT REQUIREMENT										
	SAMPLE MEASUREMENT										
	PERMIT REQUIREMENT										
	SAMPLE MEASUREMENT										
	PERMIT REQUIREMENT										

NAME/TITLE PRINCIPAL EXECUTIVE OFFICER	I CERTIFY UNDER PENALTY OF LAW THAT I HAVE PERSONALLY EXAMINED AND AM FAMILIAR WITH THE INFORMATION SUBMITTED HEREIN; AND BASED ON MY INQUIRY OF THOSE INDIVIDUALS IMMEDIATELY RESPONSIBLE FOR OBTAINING THE INFORMATION, I BELIEVE THE SUBMITTED INFORMATION IS TRUE, ACCURATE AND COMPLETE. I AM AWARE THAT THERE ARE SIGNIFICANT PENALTIES FOR SUBMITTING FALSE INFORMATION, INCLUDING THE POSSIBILITY OF FINE AND IMPRISONMENT. SEE 18 U.S.C. § 1001 AND 33 U.S.C. § 1319. (Penalties under these statutes may include fines up to $10,000 and or maximum imprisonment of between 6 months and 3 years.)	TELEPHONE	DATE
TYPED OR PRINTED		AREA CODE / NUMBER	YEAR / MO / DAY
	SIGNATURE OF PRINCIPAL EXECUTIVE OFFICER OR AUTHORIZED AGENT		

COMMENT AND EXPLANATION OF ANY VIOLATIONS (Reference all attachments here)

1. If form has been partially completed by preprinting, disregard instructions directed at entry of that information already preprinted.
2. Enter "PERMITTEE NAME/MAILING ADDRESS (and facility name/location, if different)," "PERMIT NUMBER," and "DISCHARGE NUMBER" where indicated. (A separate form is required for each discharge.)
3. Enter dates beginning and ending "MONITORING PERIOD" covered by form where indicated.
4. Enter each "PARAMETER" as specified in monitoring requirements of permit.
5. Enter "SAMPLE MEASUREMENT" data for each parameter under "QUANTITY" and "QUALITY" in units specified in permit. "AVERAGE" is normally arithmetic average (geometric average for bacterial parameters) of all sample measurements for each parameter obtained during "MONITORING PERIOD." "MAXIMUM" and "MINIMUM" are normally extreme high and low measurements obtained during "MONITORING PERIOD." (NOTE: to municipals with secondary treatment requirement, enter 30-day average of sample measurements under "AVERAGE" and enter maximum 7-day average of sample measurements obtained during monitoring period under "MAXIMUM".
6. Enter "PERMIT REQUIREMENT" for each parameter under "QUANTITY" and "QUALITY" as specified in permit.
7. Under "NO. EX" enter number of sample measurements during monitoring period that exceed maximum (and/or minimum or 7-day average as appropriate) permit requirement for each parameter. If none, enter "0".
8. Enter "FREQUENCY OF ANALYSIS" both as "SAMPLE MEASUREMENT" (actual frequency of sampling and analysis used during monitoring period) and as "PERMIT REQUIREMENT" specified in permit. (e.g., Enter "CONT." for continuous monitoring. "1/7" for one day per week. "1/30" for one day per month. "1/90" for one day per quarter, etc.)
9. Enter "SAMPLE TYPE" both as "SAMPLE MEASUREMENT" (actual sample type used during monitoring period) and as "PERMIT REQUIREMENT." (e.g., Enter "GRAB" for individual sample. "24HC" for 24-hour composite. "N/A" for continuous monitoring, etc.)

(FOLD HERE FIRST)

10. WHERE VIOLATIONS OF PERMIT REQUIREMENTS ARE REPORTED, ATTACH A BRIEF EXPLANATION TO DESCRIBE CAUSE AND CORRECTIVE ACTIONS TAKEN. REFERENCE EACH VIOLATION BY DATE.
11. If "no discharge" occurs during monitoring period, enter "NO DISCHARGE" across form in place of data entry.
12. Enter "NAME/TITLE OF PRINCIPAL EXECUTIVE OFFICER" with "SIGNATURE OF PRINCIPAL EXECUTIVE OFFICER OR AUTHORIZED AGENT." "TELEPHONE NUMBER" and "DATE" at bottom of form.
13. Mail signed Report to Office(s) by date(s) specified in permit. Retain copy for your records.
14. More detailed instructions for use of this DISCHARGE MONITORING REPORT (DMR) form may be obtained from Office(s) specified in permit.

LEGAL NOTICE

This report is required by law (33 U.S.C. 1318: 40 C.F.R. 125.27). Failure to report or failure to report truthfully can result in civil penalties not to exceed $10,000 per day of violation: or in criminal penalties not to exceed $25,000 per day of violation, or by imprisonment for not more than one year, or by both.

FOLD HERE SECOND

PLACE STAMP HERE

FOLD HERE THIRD

STAPLE HERE

RCRA REPORTING AND RECORDKEEPING

David R. Case
General Counsel
Hazardous Waste Treatment Council
Washington, D.C.

1.1 Introduction

A central purpose of the Resource Conservation and Recovery Act ("RCRA"), 42 U.S.C. §§ 6901 et seq., is to require all persons who manage hazardous wastes to maintain reports and records of their activities. RCRA directs EPA to establish numerous standards for recordkeeping and reporting, which are set forth throughout the regulations at 40 C.F.R. Parts 260-271. Congress intended for RCRA to establish a nation-wide program which, for the first time, documents the identity and quantity of hazardous wastes that are produced; tracks their shipment across the country; and records their manner and place of disposition.

In general, EPA's reporting and recordkeeping standards apply to all persons who produce hazardous wastes ("generators"); all persons who transport hazardous wastes off the site of generation ("transporters"); and all persons who treat, store, or dispose of hazardous wastes ("TSD facilities"). Even persons who do not generate hazardous wastes are indirectly affected, however, since they should maintain records of their determinations that their solid wastes are not hazardous. RCRA recordkeeping also includes manifests, contingency plans, personnel training records, notifications related to the land disposal restrictions, and much more. Reporting mainly involves the submission of comprehensive reports to EPA or a state which detail the identity, quantity, and manner of treatment and disposal of all hazardous wastes managed during the year. Reporting can also be triggered by unexpected events, such as emergency spills or unmanifested waste shipments.

Making reports and keeping records under RCRA is a big job. EPA estimates that 272 million metric tons of hazardous wastes are produced in the United States each year. All of this hazardous waste must be accounted for in RCRA records and reports. Generators, transporters, and TSD facilities must have well-planned and monitored programs to ensure accuracy and completeness, and to minimize the potential costs and burdens.

This chapter focuses on the reporting and recordkeeping duties of generators of hazardous wastes. Most of the 500,000 companies subject to RCRA requirements are generators. Transporters look to the provisions of the Hazardous Materials Transportation Act and the U.S. Department of Transportation (DOT) regulations. TSD facilities are subject to extensive recordkeeping and reporting obligations imposed by RCRA permits, which are beyond the scope of this chapter.

1.2 Overview of Standards For Generators

The challenge for generators is to adopt an efficient recordkeeping and reporting program that coordinates the diverse duties that RCRA imposes. For example, generators must determine which of their wastes are hazardous, usually by having samples of the waste tested at an analytical laboratory. Generators must then make arrangements for proper treatment and disposal of their waste, which usually involves more testing to complete the waste profile sheets required by commercial TSD facilities. In addition, the generator must provide a notification of the treatment standards and land disposal restrictions that apply to his waste, with attached analytical test results.

Thus, with respect to a single hazardous waste stream, the generator can either send the waste to an analytical laboratory three separate times, and pay three times, or he can coordinate the testing he needs at the outset. A generator should be able to obtain all required test results for a waste at one time. There are obvious advantages to a well thought-out program, not the least of which is the potential cost savings. Other opportunities to promote accuracy and efficiency can also be found, such as by using computers to maintain records such as manifests and to prepare reports. These are discussed further below.

1.3 Statutory Provisions

RCRA is surprisingly verbose about the types of records and reports that a generator must maintain. RCRA § 3002(a) states that EPA must promulgate regulations respecting--

> (1) recordkeeping practices that accurately identify the quantities of such hazardous waste generated, the constituents thereof which are significant in quantity or in potential harm to human health or the environment, and the disposition of such wastes; . . .
>
> (4) furnishing of information on the general chemical composition of such hazardous waste to persons transporting, treating, storing, or disposing of such wastes;
>
> (5) use of a manifest system . . . to assure that all such hazardous waste generated is designated for treatment, storage, or disposal in, and arrives at treatment, storage, or disposal facilities (other than facilities on the premises where the waste is generated) for which a permit has been issued . . .; and
>
> (6) submission of reports to the Administrator (or the State agency in any case in which such agency carries out a permit program pursuant to this subtitle) at least once every 2 years, setting out--
>
>> (A) the quantities and nature of hazardous waste identified or listed under this subtitle that he has generated during the year;
>>
>> (B) the disposition of all hazardous waste reported under subparagraph (A);
>>
>> (C) the efforts undertaken during the year to reduce the volume and toxicity of waste generated; and
>>
>> (D) the changes in volume and toxicity of

> waste actually achieved during the year in
> question in comparison with previous years.

Although Section 3002(a)(6), quoted above, allows reports to be filed with EPA "at least once every two years", most states have adopted annual reporting requirements. This is an important point. Almost all states now administer hazardous waste programs that have been authorized by EPA in lieu of the Federal program. See RCRA § 3006(b). States must administer programs that are equivalent to and consistent with the Federal program, but a state can also impose requirements that are more stringent that those imposed by EPA's regulations. See RCRA § 3009. Annual reporting by generators, instead of the biennial reports required by EPA's regulations, are an example of a more stringent requirement. Generators must look to state statutes and regulations for additional or more stringent reporting and recordkeeping obligations than those imposed by EPA.

1.4 Waste Analyses

The first obligation of a generator is to determine which materials he produces are wastes, and which are hazardous wastes. 40 C.F.R. § 262.11. The prescribed steps for making hazardous waste determinations are set forth in the EPA regulations. See also RCRA Hazardous Wastes Handbook, Chapter 4 (Government Institutes 8th ed.). Basically, the generator must determine whether his waste is listed by EPA, or whether it exhibits a hazardous characteristic. The generator can rely on knowledge of his raw materials, processes, and waste streams, or he can conduct analyses of the waste using the EPA-prescribed test protocols. These determinations should be in writing, supported by any relevant documentation and analytical test results.

For example, the generator may rely on the Material Safety Data Sheets ("MSDS") supplied by the producer of chemicals used in the process. The MSDS will describe the chemical and physical properties of chemical ingredients which could cause wastes to be hazardous. Or the generator can conduct periodic testing and rely on knowledge that raw materials and processes have not changed for subsequent determinations. A prudent generator will also make written determinations that secondary materials are not wastes, or that wastes are not hazardous.

A generator must then retain all records of the test results, waste analyses, and other determinations made for the purpose of complying with this regulatory duty. 40 C.F.R. § 262.40(c). Note that a generator may obtain waste analysis data for other purposes, such as during an internal environmental audit. The generator must exercise careful judgment in deciding whether any such waste analyses were "made in accordance with" the regulatory duty to determine if his waste is hazardous. If so, the record retention obligation may be triggered.

EPA requires that waste determination records be kept for at least three years from the date that the particular waste was last sent to an on-site or off-site TSD facility. 40 C.F.R. § 262.40(c). A generator should think carefully before destroying such records after the mandatory three year period, however. Waste determination records may be needed in the future to fully respond to claims for alleged unlawful disposal of hazardous wastes, liability for cleanup at Superfund sites, third-party tort actions, and other such purposes. Such claims can be asserted ten, twenty, or even fifty years after the waste has been disposed of by the generator. Without accurate records, the generator may be unable to adequately defend against such claims. Most generators retain RCRA records permanently.

Typically, a generator will periodically send a sample of his waste to an analytical laboratory for testing to determine if the waste meets a listing description or exhibits a hazardous characteristic. As discussed further below, however, the generator may later need additional test results of the same waste for other purposes. Waste analyses may be needed to satisfy DOT transportation requirements, to supply to a TSD facility, to comply with the land disposal restrictions, or for other regulatory or internal purposes. Therefore the generator should develop an overall waste analysis program which coordinates all needed analyses and minimizes duplication, costs, and related burdens. The program should also address the frequency with which periodic retesting will be done, depending on changes in raw materials, processes, or other factors that affect the waste streams.

1.5 Notification of Hazardous Waste Activity

RCRA § 3010 requires that any person who generates a hazardous waste must file a notification with EPA or an authorized state describing the location, activity, and wastes handled by that person. This should be done on EPA Form 8700-12, available from any EPA regional office.

The notification form requires only the following:
- The generator's name and address, and the name of the owner, if different;
- The name of a contact person at the facility;
- The type of regulated activity; and
- The waste codes for wastes generated.

For generators at existing facilities back in 1980, the notification should have been filed in August 1980. For new generators, the notification must be filed within 90 days of becoming subject to RCRA. For example, future regulatory actions identifying new hazardous wastes or characteristics may bring companies under RCRA regulation for the first time. These companies would file a 3010 notification within three months after EPA's regulatory action. Generators who produce and market or burn hazardous waste-derived fuel or used oil fuel also had to file a notification by February 1986.

Upon filing the notification, EPA or an authorized state will issue the generator an EPA Identification number. This ID number must be used on all reports and records required by RCRA. Generators should retain a copy of their notification form.

1.6 Manifest Records and Exception Reports

Congress created the manifest system to ensure that generators would, in many cases for the first time, track all shipments of hazardous waste to a designated TSD facility. Prior to the manifest system, it was not uncommon for generators to keep minimal or no records of their off-site shipments of hazardous waste. In addition, the manifest system compels the generator to make advanced arrangements for proper treatment and disposal at a permitted TSD facility. The hauler can no longer decide where the waste will end up.

1.6.1 The Manifest Document

Manifest requirements are set forth in 40 CFR §§ 262.20 - 262.23. Generators must use the manifest form required by the state where the TSD facility is located. If no particular form is required there, then the generator must use the form required by the state in which he is located. If neither state prescribes a form, then the generator can use any form that complies with the EPA Uniform Manifest, a copy of which is reproduced in the Appendix to 40 CFR Part 262.

The generator is responsible for completing and certifying the manifest form. Even if an outside contractor is used to prepare the manifest, such as a turn-key disposal company, the generator is still responsible under the RCRA regulations for its proper completion. An authorized employee of the generator must sign the certification on the manifest which declares, in effect, that the manifest is accurate and complete, and that the shipment has been properly classified, packed, marked, and labeled, and is "in all respects" in proper condition for transport according to the regulations. There are criminal penalties for making a "knowingly false material statement" on a manifest. See RCRA § 3008.

After the generator signs the manifest, he must obtain the signature of the transporter and retain an initial copy. The transporter must then keep the remaining copies of the manifest with the waste shipment at all times. When the waste is delivered, the designated facility must complete the tracking process by signing the manifest. The transporter then keeps a completed copy, and the designated facility must send the completed manifest back to the generator within 30 days. At that point, the generator need only retain this final manifest; the first copy can be discarded to lessen the paperwork retention burden. See 40 CFR § 262.40(a).

EPA's regulations require that the generator retain the completed manifest only for three years from the date that the waste was accepted by the transporter. Id. Because the manifest provides documentation that the waste was sent to an authorized facility, it should be retained permanently, however.

1.6.2 Exception Reports

What happens if the completed manifest is not received by the generator? EPA is concerned that until the completed manifest is returned, the generator cannot know for certain that his waste was actually received by the designated facility. Some misfortune may have befallen the waste shipment before it could reach the designated TSD facility.

Accordingly, the generator who has not received a completed manifest within 35 days of shipment must make inquiries of the transporter and TSD facility to find out what happened. 40 CFR § 262.42. The manifest may be in the mail, or it may still be sitting on someone's desk. In any event, the generator must receive a copy of the manifest bearing the handwritten signature of the TSD facility no later than 45 days after the transporter picked up the shipment. (A 100-1000 kg/mo small quantity generator has 60 days to receive confirmation.) Otherwise the generator must file an Exception Report with EPA or the authorized state.

The Exception Report consists of a copy of the original manifest and a letter of explanation. The letter must describe the efforts undertaken by the generator to locate the shipment and the results. Obviously, if the generator was unable to happily resolve the matter, EPA or the state may initiate action to locate and recover the waste. Thus the generator should make all reasonable efforts to ensure that manifests are returned to him on a timely basis, particularly in these days of expedited mail/delivery services. Even if the waste was properly received, and the manifest was just delayed, Exception Reports may be seen as reflecting poorly on a generator's compliance.

Copies of Exception Reports must also be retained for at least three years from the due date of the report. 40 CFR 262.40(b).

1.6.3 Use of Computers

There are many commercially available software packages that allow a generator to use computer systems for tracking waste shipments and preparing manifests. Software vendors usually advertize in the popular trade publications. In addition, some companies have found it cost-effective to have software written or adapted for their particular situation.

A computerized manifest system will typically keep track of waste shipments and generate manifests directly from the computer. The program can provide warnings when stored wastes are approaching the ninety day limit at a generator's facility, and also when manifests have not been returned by the TSD facility within a defined period. The data base of waste shipments in the computer can then be used to produce annual reports. The computer also provides long-term storage of the data, although this cannot substitute for retaining the actual signed manifests received from the TSD facilities.

1.7 Ninety-Day Storage Facility Records

A generator may store his own hazardous wastes on-site for up to 90 days without a RCRA permit, provided the requirements of 40 CFR 262.34 are met. This regulations, in turn, incorporates many, but not all, of the standards that would apply to an interim status storage facility. Among these requirements are a number of recordkeeping provisions.

1.7.1 Contingency Plans

Specifically, the generator must prepare a written Contingency Plan that sets out the response measures to be taken in the event of a fire, explosion, or other unplanned release of hazardous wastes. The Contingency Plan must be maintained at the generator's facility and be available for inspection. Whenever the Contingency Plan is implemented, the Emergency Coordinator must submit a report to EPA within 15 days of the incident. The report must describe the incident, the hazardous wastes involved, the extent of any injuries, an assessment of the actual or potential hazardous posed, and the quantity/disposition of any recovered waste materials.

1.7.2 Training Records

The generator must also develop an employee training program, which focuses on emergency preparedness and response. Training records must be kept. These records must include the job title and description for each person involved in the hazardous waste management program, a written description of the type and amount of training to be given to each person, and documentation of the training actually given. Current records must be maintained at the facility.

1.7.3 Inspection Records

Generators must also comply with the interim status standards for containers and tanks. Among these standards are the requirements for regular inspections by the generator's personnel of the container and tank storage areas. Although the regulations at 40 CFR 265, Subparts I and J, do not require that written records of the inspections be maintained, it is advisable to do so. The generator should prepare a written inspection plan which describes the nature and frequency of these self-inspections, including a checklist of items to look for during the inspection. A written log showing that the inspections actually took place should also be maintained. See 40 CFR § 265.15 for guidance.

1.8 Land Ban Notifications and Certifications

Generators play a key role in the land disposal ban program, and as a result additional records are required. Very soon all hazardous wastes will be subject to restrictions on land disposal. That means that generators must prepare certain documents for every shipment of hazardous wastes that are subject to the land ban regulations. This can be a formidable task.

1.8.1 Notifications

First of all, the generator must determine which treatments standards and land disposal prohibitions apply to his waste. 40 CFR § 268.7. He can do this by testing his waste or a waste extract, as required, or by using his general knowledge of the waste. The required test methods are specified in 40 CFR Part 268. Generally, a total waste analysis is required when the BDAT treatment standard is based on a destruction or removal technology. An analysis of a TCLP waste extract is required when BDAT is based on stabilization, and for the solvent/dioxin wastes. Where both types of technologies were used for BDAT, then both types of waste analyses must be used. If the generator uses his general knowledge, all supporting data relied on to make this determination must be maintained on-site in the generator's files.

With each shipment of hazardous waste that is subject to land disposal restrictions, the generator must prepare a Notification of the applicable treatment standards. No form is prescribed. Many TSD facilities supply forms to their customers. The Notification must state:

- The EPA Hazardous Waste Number (e.g., F001 for certain listed spent solvents; D001 for ignitable wastes) for each waste;

- The BDAT treatment standards and all applicable prohibitions for each waste (note that the actual treatment standards must be stated, not just a reference to the regulation where the treatment standard appears); and

- The manifest number associated with the waste.

In addition, all waste analysis data that is available for the waste must be attached to the Notification. Currently, EPA requires all such waste analyses to accompany every Notification, even if previously supplied to the TSD facility.

Even if the generator sends the same waste to the same TSD facility on a regular basis, a full Notification must accompany each shipment. EPA may relax this paperwork burden in the future. If the generator's waste is subject to a national capacity variance, a no-migration exemption, or a case-by-case extension, he must forward a notice with the waste to the land disposal facility receiving his waste, stating that the waste is exempt from the land disposal restrictions.

The purpose of the Notification, of course, is to ensure that the treatment facility that receives the waste is fully apprised of the treatment standards that apply. The treatment facility must, in turn, certify to the land disposal facility that the waste has been treated to meet the required treatment standards before the waste residue may be land disposed. It may be necessary for more than one treatment facility to treat the waste; e.g., the first facility may remove inorganic constituents and then send the partially treated waste to a second facility where the organic constituents are destroyed. In that case, the generator's Notification is used by the first facility to notify the second facility of all applicable treatment standards. The second treatment facility will make the required certification to the land disposal facility.

1.8.2 Certifications

It may be that the waste as generated already meets the treatment standards and may be directly land disposed. In that case, the generator must prepare both the Notification and the required Certification. The Certification must include the following:

- The EPA Hazardous Waste Number;
- The corresponding treatment standards and applicable prohibitions;
- The manifest number; and
- Waste analysis data, where available.

In addition, the Certification must be signed by an authorized representative of the generator to the following effect:

> I certify under penalty of law that I personally have examined and am familiar with the waste through analysis and testing or through knowledge of the waste to support

this certification that the waste complies with the treatment standards specified in 40 CFR Part 268 Subpart D and all applicable prohibitions set forth in 40 CFR 268.32 or RCRA section 3004(d). I believe that the information I submitted is true, accurate and complete. I am aware that there are significant penalties for submitting a false certification, including the possibility of a fine and imprisonment.

The generator is not required to keep copies of the Notifications and Certifications, which are retained at the disposal facility. Generators should consider retaining copies, however, for their own records.

1.9 Annual or Biennial Reports

Finally, generators must submit reports to EPA or an authorized state that comprehensively describe the hazardous waste activities at the facility. 40 CFR § 262.41. EPA requires biennial reports on March 1 of even-numbered years for the previous odd-numbered calendar year. Many states require annual reports.

The report should be submitted on EPA Form 8700-13A. Besides the generator's name, location, and ID number, the report must identify all transporters and TSD facilities that were used. The report also calls for a description, EPA Hazardous Waste Code, DOT hazard class, and quantity of all hazardous wastes shipped off site.

The report is also used to document the generator's waste minimization program. The generator must describe the efforts undertaken during the year to reduce the volume and toxicity of wastes generated. The report must then document the changes in volume and toxicity actually achieved during the year in comparison to previous years. These annual or biennial reports must be retained for at least three years.

THE CLEAN AIR ACT: R&R REQUIREMENTS

G. Vinson Hellwig
Vice President, Air Division
TRC Environmental Consultants, Inc.

1.0 INTRODUCTION

Under the Clean Air Act (CAA) Amendments of 1977, recordkeeping and reporting are required for sources subject to New Source Performance Standards (NSPS), National Emissions Standards for Hazardous Air Pollutants (NESHAPs), Prevention of Significant Deterioration (PSD), New Source Review (NSR), compliance orders and specific permit conditions. Each of these areas has separate recordkeeping and reporting requirements, and we will review what those requirements are and whether they are subject to federal, state or local agency requirements.

Recordkeeping and reporting requirements are basically used by the agencies as an enforcement tool to document continuing compliance. The documentation must be as accurate as possible; tampering with any type of reporting or recordkeeping is not permitted and can result in legal prosecution.

2.0 NSPS AND NESHAP REQUIREMENTS

NSPS and NESHAPs are covered under Section 111 and Section 112, respectively, of the CAA Amendments of 1977.

2.1 NSPS

Section 111 of the CAA provides for the development and implementation of regulations under New Source Performance Standards (NSPS). These standards establish certain categories of air pollution sources which, after proper notification to the public, require that emission standards be established. As part of these NSPS, certain recordkeeping requirements are implemented to assure that the standards are being met.

Previously promulgated standards are found in the Code of Federal Regulations, Title 40, Part 60 (40CFR60). The Code of Federal Regulations is published each July, with any NSPS promulgated since the previous July included. 40CFR60 sets out the requirements and general provisions as they apply to the owner/operator of any stationary source which contains an affected facility. Any facility where construction or modification began after the date of proposal in the Code of Federal Regulations will be subject to the standard listed. NSPS promulgated after July will appear in the Federal Register (FR); potentially affected facilities must review the FR for any applicable NSPS revisions.

Most of the actual compliance and enforcement for NSPS has been delegated to the state agencies at the present time, while the EPA has reserved some authority.

There is a general provision under 40CFR60.7, "Notification and Recordkeeping", which deals with the installation and operation of Continuous Emission Monitoring Systems (CEMS). Any owner/operator that is required to have a CEMS must meet specific reporting requirements. Additional monitoring requirements are found as part of 40CFR60.13, and other reporting requirements are found under the individual NSPS categories.

2.1.1 CEMS Requirements

NSPS require that any owner/operator subject to the provisions of mandatory continuous emission monitoring maintain those operating records, as well as records of any startup, shutdown or malfunction of an affected facility's air pollution control equipment during any period in which a continuous monitoring system or device is inoperative. In addition, every source that is required to install a CEMS must submit a written report of excess emissions, as defined in applicable portions of the regulations. These reports must be submitted every calendar quarter, and they must be postmarked by the 30th day of the end of each calendar quarter. The report must include certain information:

1. The magnitude of the excess emissions computed in accordance with 40CFR60.13(h), any conversion factors used, the date and time of the commencement and completion of each time period of excess emissions.

2. Specific identification of each period of excess emissions that occurs during the startup/shutdown malfunctions, the nature and cause of the malfunction, and, if known, the corrective actions that took place.

3. The date and time identifying each period when the continuous monitoring system was inoperative, except for zero/span checks, and the nature of repairs or adjustments.

4. If no excess emissions have occurred or the monitors have not been inoperative, repaired or adjusted, the report should state this.

The owner/operator subject to these provisions has to maintain a file of all measurements, source testing, CEMS performance evaluations, calibration checks, adjustments and maintenance performed on the system, and any other information required under a specific Subpart of the regulations. If subject to specific Subsection, these files must be retained for at least two (2) years following the date any such measurements are reported.

All CEMS must meet the performance specifications as listed under 40CFR60 Appendix B. If they are used to demonstrate compliance with emission limits on a continuous basis, they are also subject to 40CFR60 Appendix F.

SUBPARTS THAT ARE SUBJECT TO SPECIFIC CEMS RECORDKEEPING AND REPORTING REQUIREMENTS

Subpart D	Fossil Fueled Fired Steam Generators
Subpart Da	Electric Utility Steam Generating Units
Subpart Db	Industrial Commercial Institutional Steam Generating Units
Subpart F	Portland Cement Plants
Subpart G	Nitric Acid Plants
Subpart H	Sulfuric Acid Plants
Subpart J	Petroleum Refineries
Subpart N	Primary Emissions from Basic Oxygen Process Furnaces, Construction Commenced after June 11, 1973
Subpart Na	Secondary Emissions from Basic Oxygen Process Steel Making Facilities, Construction Commenced after January 20, 1983
Subpart O	Sewage Treatment Plants
Subpart P	Primary Copper Smelters
Subpart Q	Primary Zinc Smelters
Subpart R	Primary Lead Smelters
Subpart S	Primary Limited Reduction Plants
Subpart T	Phosphate Fertilizer Industry: Wet Process Phosphoric Acid Plants
Subpart U	Phosphate Fertilizer Industry: Super Phosphoric Acid Plants
Subpart V	Phosphate Fertilizer Industry: Dye Ammonium Phosphate Plants
Subpart W	Phosphate Fertilizer Industry: Triple Super Phosphate Plants

Subpart Y	Coal Preparation Plants
Subpart Z	Ferroalloy Production Facilities
Subpart AA	Steel Plants: Electric Arc Furnaces Constructed after October 21, 1974 and on or before August 17, 1983
Subpart AAa	Steel Plants: Electric Arc Furnaces in Argon Oxygen Decarbonization Vessels Constructed after August 7, 1983.
Subpart BB	Performance for Kraft Pulp Mills
Subpart CC	Glass Manufacturing Plants
Subpart EE	Surface Coating of Metal Furniture
Subpart GG	Stationary Gas Turbines
Subpart HH	Lime Manufacturing Plants
Subpart KK	Lead-Acid Battery Manufacturing Plants
Subpart LL	Metallic Mineral Processing Plants
Subpart MM	Automobile Light Duty Trucks Surface Coating Operations
Subpart NN	Phosphate Rock Plants
Subpart PP	Ammonium Sulfate Manufacture
Subpart QQ	Graphics Arts Industry: Publication Rotogravure Printing
Subpart RR	Pressure Sensitive Tape and Label Surface Coating Operations
Subpart SS	Industrial Surface Coating: Large Appliances
Subpart TT	Metal Coil Surface Coating
Subpart UU	Asphalt Processing and Asphalt Roofing Manufacture
Subpart VV	Equipment Leaks of VOC in the Synthetic Organic Chemical Manufacturing Industry
Subpart WW	Beverage Can Surface Coating Industry
Subpart BBB	Rubber Tire Manufacturing Industry
Subpart FFF	Flexible Vinyl and Urethane Coating and Printing
Subpart GGG	Equipment Leaks of VOC in Petroleum Refineries
Subpart HHH	Synthetic Fiber Production Facilities
Subpart LLL	Onshore Natural Gas Processing: SO_2 Emission
Subpart OOO	Non-Metallic Mineral Processing Plants
Subpart PPP	Wool Fiberglass Insulation Manufacturing Plants
Subpart QQQ	Petroleum Refinery Wastewater Systems
Subpart RRR	Magnetic Tape Coating Facilities

2.1.2 Other Recordkeeping and Reporting Requirements

Many categories of NSPS require keeping other records. Most of these deal with production records; however, in the area of coating and volatile organic compound (VOC) control, the requirements for records and tracking are far more complex. These deal with the quantity of solvent and calculation of solvent usage in estimating what is emitted into the atmosphere.

There are other reporting requirements that deal with VOC recordkeeping. For example, Subpart K and Subpart Ka, referring to VOC storage vessels, require that data be maintained on the Reid vapor pressure, and the quantity of material that is retained in these vessels.

There are a number of NSPS standards that require process monitoring and recordkeeping; for instance, the phosphate fertilizer industry has several NSPS that it is subject to, each with specific recordkeeping requirements. As stated earlier, all of these records must be maintained for a minimum of two (2) years. The recordkeeping and reporting requirements for leaks in the synthetic organic chemical manufacturing industry and petroleum refineries are very specific as to what types of records must be maintained when performing leak checks from the various points within the refinery operations such as valves, seals, pumps, etc. Although this is not a continuous monitoring activity, it is an ongoing program for which records must be maintained at any of the facilities that are subject to these regulations.

A company that is subject to NSPS should carefully review the applicable regulations and determine what records must be maintained and what must be reported to the agencies.

2.2 NESHAPs

Section 112 of the CAA deals with the National Emissions Standards for Hazardous Air Pollutants (NESHAPs). This portion of the Act requires EPA to promulgate standards and regulatory requirements for sources that emit pollutants identified and classified as hazardous air pollutants. These pollutants include such chemicals as beryllium, mercury, asbestos, vinyl chloride, benzene, Radon 222 from underground uranium mines, thermonuclide emissions from Department of Energy facilities, thermonuclide emissions from all metal phosphorus facilities, Radon 222 emissions from licensed uranium mill tailings, arsenic emissions from glass manufacturing plants, arsenic emissions from primary copper smelters, and fugitive emissions sources from equipment leaks.

Specific pollutants that are regulated are:

Asbestos
Benzene
Beryllium
Coke Oven Emissions
Inorganic Arsenic
Mercury
Radionuclides
Vinyl Chloride

Other sources that have been identified as potential listings for NESHAPs and are under consideration at the present time are acrylonitrile, 1,3-butadiene, cadmium, carbon tetrachloride, chlorinated benzenes, chlorofluorocarbon 113, chloroform, chloroprene, chromium, copper, epichlorohydrin, ethylene dichloride, ethylene oxide, hexachlorocyclopentadiene, manganese, methyl chloroform, methylene, nickel, perchloroethylene, phenol, polycylic organic matter, toluene, trichloroethylene, vinyldene chloride, zinc and zinc oxide.

Previously promulgated standards are found in the Code of Federal Regulations, Title 40, Part 61 (40CFR61). The Code of Federal Regulations is published each July, with any NSPS promulgated since the previous July included. 40CFR61 sets out the requirements and general provisions as they apply to the owner/operator of any stationary source which contains an affected facility. Any facility where construction or modification began after the date of proposal in the Code of Federal Regulations will be subject to the standard listed. NESHAPs promulgated after July will appear in the Federal Register (FR); potentially affected facilities must review the FR for NESHAP revisions.

Most of the actual compliance and enforcement for NESHAPs has been delegated to the state agencies at the present time, while the EPA has reserved some authority.

There is a general provision under 40CFR61.14, "Monitoring Requirements", which deals with the installation and operation of Monitoring Systems (MS). Any owner/operator that is required to have an MS must meet specific reporting requirements. Other reporting requirements are found under the individual NESHAP categories.

For purposes of discussion, we will deal with the reporting requirements of currently regulated pollutants. Since there are very few uranium producers, users of beryllium, and coke oven operators, we will not discuss those areas.

Mercury is a regulated pollutant and is subject to monitoring and reporting requirements for various processes. These include mercury processing, chloroalkali plants, alkali metal hydroxide facilities and dry sewer sludge incineration facilities. Reporting is required for both continuous emission monitoring and other related indirect emission monitoring. Indirect emission monitoring refers to monitoring of pH alkali, pH of liquid scrubbers, liquid flow rates and exit gas temperatures. Vinyl chloride reporting and recordkeeping requirements deal with both continuous emission releases and emergency releases and leaks. The leak monitoring requirements are very similar to those required for leaks in the refinery industry.

Subpart J of the NESHAP standards deal with fugitive emission leaks from benzene. The owner/operator must maintain records from leaks from pumps, compressors, relief valves, etc. for a period of two (2) years.

Metal phosphorus plants under Subpart K have monitoring and recordkeeping requirements on the control devices, such as installation date, calibrations, etc. It also requires information on continuous emission monitoring systems including monitoring of primary and secondary current and voltage in the electric fields that go to the electric furnaces that produce the phosphorus.

Subpart M of NESHAP standards deals with asbestos. Any owner/operator must maintain certain information on their control equipment and disposal of collected asbestos. This applies to sources that produce and actually use the asbestos in a manufacturing operation. For firms dealing with asbestos removal or disposal in demolition or renovation situations, there is a different set of reporting requirements that must be met, and these are subject to a separate Act, the Asbestos Hazard Emergency Response Act (AHERA). Those particular asbestos regulations and requirements should be reviewed.

Under Subpart N, inorganic arsenic emissions from glass manufacturing plants are subject to monitoring reporting requirements that include requirements for continuous emission monitors and other recordkeeping. The recordkeeping requirements are very specific as to what records must be maintained on production and control equipment operation. Likewise, Subpart O has monitoring requirements for arsenic emissions from primary copper smelters.

3.0 PSD and NSR

Prevention of Significant Deterioration (PSD) and New Source Review (NSR) are specific requirements of Part C of the Clean Air Act that deal with sources installed in areas classified as attainment for ambient air quality.

Prevention of Significant Deterioration is a federal program that has been delegated to most state agencies, but is still under the purview of the U.S. EPA. New Source Review is a new program that is administered by state agencies. The requirements by which facilities are subject to New Source Review vary from state to state. This is a program that is reviewed but not administered by EPA, and deals with sources that are smaller than those subject to PSD.

Monitoring and recordkeeping for PSD can begin prior to the actual construction of a facility. For many facilities in certain areas of the country, there are PSD ambient air monitoring requirements. These requirements may include monitoring for ambient air quality as well as for meteorological data. In the case where meteorological data is not available for a period of two years and at close enough proximity to the facility, the proposed facility must monitor ambient air quality data, and this data must meet certain criteria requirements specified in the EPA Guidelines. Likewise, where existing ambient air quality data is not available, one year of ambient air quality data for the regulated pollutant or pollutants in question must be obtained and must meet certain requirements. An example of this would be monitoring for NO_x or SO_2 in an area where there is no monitoring data, and the facility is going to be a major combustion source of those pollutants.

New Source Review pre-construction monitoring requirements can vary from state to state. Typically, states tend to be more lenient with NSR than with PSD, because NSR deals with smaller sized facilities which tend to have less impact on ambient air quality.

3.1 Permit Requirements

PSD and NSR permits are required under the CAA. These permitting requirements can go beyond the current recordkeeping and reporting requirements that may otherwise apply under the CAA. It is not uncommon for a state or the U.S. EPA to include monitoring or reporting requirements above and beyond the minimum required for such a facility under NSPS. For instance, a facility that is going to combust sewage sludge has certain monitoring requirements on the scrubber. The agency may put additional requirements above and beyond simply recording the pressure drop on that scrubber, such as requiring liquid flow measurements on the scrubber or temperature monitoring on the incinerator.

Likewise, a state agency under NSR can put additional permit requirements on sources. The agency typically selects regulatory requirements that would be required of an NSPS facility, even though the facility in question is not subject to NSPS or PSD regulations. For instance, an agency might require continuous emission monitoring or some other type of direct or indirect monitoring such as pH on the scrubber liquor or continuous emissions monitoring requirements on a combustion source for opacity, NO_x, CO, or oxygen.

4.0 COMPLIANCE ORDERS

Compliance orders issued by the EPA or a state agency typically have recordkeeping and recording requirements. These are negotiable items, and may be dictated by the state or EPA. Typically, these are negotiated with the company after a notice of violation (NOV) or non-compliance has been issued, but before an administrative or judicial order may be issued. The conditions that are part of the administrative or judicial order are typically negotiated so that the reporting and recordkeeping requirements tend to be burdensome on the source. One issue to raise in the situation where there are excessive reporting requirements is whether the agency imposing the reporting requirements has the staff or the resources to evaluate the information being submitted. This is a point that should be made if the agency is requiring all records for continuous emission monitoring on a combustion source to be submitted for pollutants such as oxygen, CO, CO_2, NO_x, SO_2 and opacity. Are they indeed going to evaluate all the data the source is submitting? Are they putting undue requirements on that source?

4.1 Direct Monitoring

Direct monitoring is where monitoring occurs for a specific pollutant or diluent. Specific pollutants that are monitored are typically regulated pollutants, for instance, oxides of sulfur, oxides of nitrogen, carbon monoxide, and opacity. These are regulated limits that are measured. Diluents that are related to the emissions in a combustion source are CO_2 and O_2. All of these are direct monitoring; in other words the monitoring that takes place relates directly to emission limits.

4.2 Indirect Monitoring

Indirect or parameter monitoring deals with other items that are related to, but do not concern actual monitoring of pollutants. Some examples of these are temperature monitoring (reporting and recordkeeping requirements on certain incinerators to be sure that the proper combustion temperatures are maintained); measurement of scrubber liquor where hypochlorite or a caustic scrubber is used; pressure drop across a scrubber; records on electrostatic precipitator performance; maintenance records or records of voltage or current drops, sparks per minute, etc.

4.3 Reporting

The compliance order may have a variety of reporting requirements to the agencies. The simplest of these would be typically where the instruments, either directly or indirectly, indicate there was a violation that occurred, and the duration, date, time, corrective action taken, etc. must be reported to the agency. Some agencies may actually request the reporting be done of all operations or monitoring. This again may be the issue of whether or not the agency has the capacity or manpower to actually evaluate such lengthy reports. Recordkeeping typically would be required above and beyond the malfunction stage, so that the agency could verify upon an audit or inspection that indeed the owner/operator was keeping adequate records. These records would back up what was stated as far as the violation is concerned; that type of recordkeeping is typically required in a compliance order.

5.0 PERMIT REQUIREMENTS

Permits issued by a state or local agency are authorized under the Clean Air Act, may become part of the State Implementation Plan, and any permit requirements can be federally enforceable. If a state agency has made specific permit requirements a part of their state implementation plan, these can be required and enforced as discussed in Section 4.

5.1 Negotiation

As was discussed earlier, permit conditions as with conditions under compliance orders are negotiable, or open to discussion with the state agency. In some cases, negotiations may be a give and take affair, with the facility giving up certain other points on the permit process, such as hours of operation or emission limits. It is up to the compliance source to determine how they can work with the permit conditions. Quite often there is an economic trade-off, when monitoring might be more expensive than limiting hours of operation. In most cases, unless they are explicit in the regulations, permit requirements are a negotiable item with the agencies.

REPORTING AND RECORDKEEPING
REQUIREMENTS UNDER
THE COMPREHENSIVE ENVIRONMENTAL RESPONSE,
COMPENSATION, AND LIABILITY ACT OF 1980

THEODORE W. FIRETOG
Environmental Counsel
Shea & Gould

I. INTRODUCTION

The Comprehensive Environmental Response, Compensation, and Liability Act of 1980, 42 U.S.C. §9601 et seq., commonly referred to as "Superfund" or "CERCLA", was passed by Congress and signed into law on December 11, 1980. The primary purpose of CERCLA is to provide the funding and authority by which the federal government can effectively respond to the uncontrolled release of hazardous substances from any vessel or facility and to provide the mechanism for cleaning up the hundreds of inactive hazardous waste disposal sites around the country. CERCLA accomplishes its goal not through the use of extensive regulations but by imposing strict cleanup and emergency reporting requirements on a broad class of responsible parties.

The emergency notification requirements of CERCLA are contained in Section 103(a) of the statute. These requirements will be described in Part II of this Outline. Other reporting requirements of Section 103 as well as those which may be required by specific administrative orders issued pursuant to Section 104 of CERCLA, will be discussed in Part III.

II. EMERGENCY NOTIFICATION REQUIREMENTS (SECTION 103(a))

Although CERCLA does not require any type of periodic reporting, it does contain certain specific notification requirements. Section 103(a) of CERCLA requires certain parties to immediately notify the National Response Center [at (800) 424-8802 (in the Washington D.C. at area (202) 267-2675)] of an unpermitted release to the environment of a reportable quantity of a hazardous substance from a vessel or an offshore or an onshore facility. The National Response Center, in turn, will then notify all appropriate agencies.

A. Parties Subject to Emergency Notification

The notification requirements of Section 103(a) apply to any person in charge of a vessel or facility from which a hazardous substance has been released to the environment in a reportable quantity. Although the phrase "person in charge" is not defined in CERCLA, presumably it is more inclusive than those persons defined as "owners" or "operators".

The term "person" is defined broadly in the statute to include individuals, firms, corporations, associations, partnerships, consortiums, joint ventures, commercial entities, and governmental bodies.

The Section 103(a) reporting requirement arises as soon as the person in charge of a vessel or facility receives actual knowledge of such a release. However, for the reporting requirement to arise, there must be a release involving a hazardous substance, in a reportable quantity.

B. Releases Requiring Notification

The notification requirements of Section 103(a) of CERCLA are triggered once the person in charge of a facility receives actual knowledge of the release of a reportable quantity of a hazardous substance into the environment. The term "release" is broadly defined in CERCLA to include:

> any spilling, leaking, pumping, pouring, emitting, emptying, discharging, injecting, escaping, leaching, dumping, or disposing into the environment (including the abandonment or discarding of barrels, containers, and other closed receptacles containing any hazardous substance or pollutant or contaminant) . . . .

The term "release" does not include the following:

> (1) any release which results in exposure to

persons solely within a workplace, with respect to a claim which such persons may assert against the employer of such persons;

(2) emissions from the engine exhausts of a motor vehicle, rolling stock, aircraft, vessel, or pipeline pumping station engine;

(3) release of source, by-product, or special nuclear material from a nuclear incident, as those terms are defined in the Atomic Energy Act of 1954;

(4) release of source, by-product, or special nuclear material from any processing site designated under the Uranium Mill Tailings Radiation Control Act of 1978; and

(5) the normal application of fertilizer.

Furthermore, federally permitted releases do not have to be reported. The term "federally permitted release" is defined in Section 101(10) of CERCLA to mean generally any discharge that is in compliance with a permit issued pursuant to the Federal Pollution Control Act, the Resource Conservation and Recovery Act, the Marine Protection, Research and Sanctuaries Act of 1972, the underground injection control program established under the Safe Drinking Water Act, the Clean Air Act, and the Clean Water Act, as well as any discharge authorized under the Atomic Energy Act. This "federally permitted release" exemption to the reporting requirements of Section 103(a) applies whether or not the permit is issued by a federal, state, or local authority. If the release, however, is not covered or violates the conditions of the permit, the release may have to be reported depending on whether the release involved a hazardous substance in a reportable quantity.

C. <u>Hazardous Substance Requirement</u>

Only unpermitted releases into the environment of a hazardous substance must be reported. Generally, "hazardous substance" is defined in CERCLA to include the following:

> (A) any substance designated pursuant to section 1321(b)(2)(A) of Title 33, (B) any element, compound, mixture, solution, or substance designated pursuant to section 3001 of the Solid Waste Disposal Act [42 U.S.C.A. §6921] (but not including any waste the regulation of which under the Solid Waste Disposal Act [42 U.S.C.A. §6901 et seq.] has been suspended by Act of Congress), (D) any toxic pollutant listed under section 1317(a) of Title 33, (E) any hazardous air pollutant listed under section 112 of the Clean Air Act [42 U.S.C.A. §7412], and (F) any imminently hazardous

> chemical substance or
> mixture with respect to
> which the Administrator has
> taken action pursuant to
> section 2606 of Title 15.

The term "hazardous substance" does not include petroleum, including crude oil, natural gas, natural gas liquids, liquefied natural gas, or synthetic gas usable for fuel (or mixtures of natural gas and such synthetic gas).

A composite list of hazardous substances is contained in 40 CFR part 320. As of October 1989 there were 724 substances on the hazardous substances CERCLA list.

There may be instances where a substance, not listed as a CERCLA hazardous substance, is released into the environment and upon its release forms a listed hazardous substance. In such instances, the emergency notification requirements of Section 103(a) would be triggered if the hazardous substance formed as a result of the release equals or exceeds the reportable quantity for that hazardous substance.

D. <u>Reportable Quantities</u>

An unpermitted release of a hazardous substance into the environment must be reported to the National Response Center when the person in charge of the facility has received actual knowledge of such a release, but only if the release was of a reportable quantity. EPA's regulations at 40 CFR Part 302 not only contains the composite list of hazardous substances subject to the reporting requirement, but also sets forth the reportable quantities for each substance.

Suppose, for example, there was an unpermitted release of 11 pounds of ammonium picrate into the environment from your facility. By referring to Table 302.4 in Part 302 of the regulations, you would be able to determine that the reportable quantity for that substance is 10 pounds or 4.54 kilograms. Thus, the reporting requirements of Section 103(a) of CERCLA may have been triggered. Incidentally, Table 302.4 also contains additional useful information. Again referring to ammonium picrate, you can see that the table identifies its Chemical Abstract Service Registry Number (CASRN) as 131748, and lists the common regulatory synonyms for the substance (other names by which the substance is identified in other statutes), which include Phenol, 2, 4, 6-trinito-, and ammonium salt. Also because ammonium picrate is a listed waste under RCRA, the table includes its RCRA waste number, P009.

If in the example above, 11 pounds of a mixture containing 3 pounds of ammonium picrate and 8 pounds a non-hazardous substance were released, no CERCLA notification would be required. For mixtures or solutions, a reportable release occurs where a component hazardous substance is released in a quantity equal to or greater than its reportable quantity.

In determining whether a release or a reportable quantity of a hazardous substance has occurred, a period of 24 hours is used for measuring the quantity released.

Section 103(f)(2) sets out specific requirements for reporting certain continuous releases. No notification is required if the release is continuous and stable in quantity and rate, and the release had already been reported. Notification must still be given annually and when there is a statistically significant increase in the quantity of any hazardous substance released, above that previously reported or occurring.

Certain hazardous substances are not listed in Table 302.4. These substances include RCRA hazardous wastes exhibiting the characteristics of ignitability, corrosivity, or reactivity. For these substances, a reportable quantity of 100 pounds has been established. For RCRA wastes that exhibit extraction procedure (EP) toxicity, the reportable quantity is that which is listed in Table 302.4 for the contaminant on which the characteristic of EP toxicity is based.

E. **Local Publication of Release Information**

Section 111(g) of CERCLA requires the owner and operator of any vessel or facility from which a hazardous substance has been released to provide reasonable notice to potential injured parties by publication of such notice in local newspapers serving the affected area. This notice requirement remains in effect until EPA promulgates specific rules and regulations covering the scope and form of the required notice.

F. **Penalties**

Section 103(b) of CERCLA provides for criminal penalties which may be imposed on any person in charge of facility who fails to notify the National Response Center (or other appropriate agency) pursuant to Section 103 of CERCLA of a release of a hazardous substance as soon as that person has knowledge of such a release or who submits any information in the notification which that person knows to be false or misleading. Such a person, upon conviction, may be fined (in accordance with title 18 of the United States Code) up to $250,000 ($500,000 for a corporation) or imprisoned up to 3 years (or not more than 5 years in the case of a second or subsequent conviction), or both. The information obtained from such notification, however, cannot be used by the government against the person providing such notice or information in any criminal case, except in a prosecution for perjury or for giving a false statement.

Section 109 of CERCLA provides for administrative civil penalties that may be imposed for violations of the statute, including those relating to CERCLA's emergency notification requirements. There are two categories of administrative civil penalties which may be assessed for violations of the Section 103(a) of CERCLA: a Class I administrative penalty of not more than $25,000 per violation, or a Class II administrative penalty of not more than $25,000 per day for each day during which the violation continues. In the case of a second or subsequent violation the amount of the Class II penalty may be increased to not more than $75,000 for each day during which the violation continues.

In the case of a Class I penalty, no penalty may be assessed unless the person accused of the violation is given notice and an opportunity for a hearing with respect to the violation. In determining the amount of the penalty, EPA must take into account the nature, circumstances, extent and gravity of the violation and the violator's ability to pay, any prior history of such violations, the degree of culpability, and any economic benefit or savings which resulted from the violation.

Class II administrative penalties must be assessed and collected after notice and an opportunity for a hearing on the record in accordance with the procedures set forth in the Administrative Procedures Act, 5 U.S.C. Section 554. This is a more formal process than that which is required for Class I assessments.

Actions for judicial assessments of civil penalties may also be brought in United States district courts. Such judicial assessments carry the same penalties as Class II administrative actions.

III. OTHER CERCLA REPORTING REQUIREMENTS

In addition to the emergency notification requirements of Section 103(a) of CERCLA, Section 103 contains a one-time notice provision. Under Section 103(c), any person who owned or operated a facility at which hazardous substances were stored, treated or disposed of, and such a facility had not been issued a permit under, or had been accorded interim status under RCRA, must have notified EPA by June 9, 1981, of the existence of the facility, specifying the amount and type of any hazardous substance to be found there, and any known, suspected, or likely releases of such substances. Pursuant to Section 103(d), records are to be kept on such facilities for a period of fifty years beginning with

December 11, 1980 or the date of establishment of such records (whichever is later).

Section 104 of CERCLA provides the authority by which the government may respond to the release or to the substantial threat of a release into the environment of any hazardous substance or of any pollutant or contaminant which may present an imminent and substantial danger to the public health or welfare. Pursuant to Sections 104(b) and 104(e)(1) the Administrator may undertake such investigations and other information gathering as he may deem necessary or appropriate to identify the existence and extent of the release or threat of release or the need for response actions. In exercising the Administrators authority under these Sections, it is EPA's position that the Administrator can issue specific compliance orders which would require the reporting of such information. Such information is not required to be submitted without the issuance of such orders.

OSHA Recordkeeping and Reporting Requirements

by Robert D. Moran
Cooter & Gell
Washington, D.C.

OSHA recordkeeping and reporting requirements are covered in section 8(c), 29 U.S.C. §657(c), of the Occupational Safety and Health Act of 1970. Regulations implementing that section of the act are codified in Part 1904 of Title 29, Code of Federal Regulations, but there are also many OSHA standards that include recordkeeping requirements. For example, 24 CFR §1910.134(e)(3) requires that there shall be a written record kept of the procedures "covering safe use of respirators in dangerous atmospheres that might be encountered in normal operations or in emergencies", and 29 CFR §1910.95(m) requires employers to maintain an accurate record of all noise level measurements that are required to be made by OSHA's Hearing Conservation Standard.

The Reasons for OSHA Recordkeeping

OSHA standards are binding regulations that employers must observe in order to maintain a safe and healthful workplace. They are intended to be the "heart" of the OSH Act. By the same token, a systematic recordkeeping program would be the "brains" of the Act. Congress recognized that the government must first know the particulars of the job injury and illness problem before it could begin to solve it. Records would be developed to provide that knowledge.

The Senate Committee that reported the legislation stated that: "[A]n essential first action under this bill should be the institution of adequate statistical programs." That is true because:

> "Full and accurate information is a fundamental precondition for meaningful administration of an occupational safety and health program. At the present time, however, the Federal government and most of the states have inadequate information on the incidence, nature, or causes of occupational injuries, illnesses, and deaths."[1]

1/ Legislative History of the Occupational Safety and Health Act of 1970 at p. 156, Committee on Labor and Public Welfare, United States Senate, 92nd Congress, 1st Session. This publication will be referred to hereafter simply as "Legislative History." The House Committee Report contained similar statements.
See Legislative History at page 860.

The Act granted OSHA full authority to do whatever was necessary in order to compile the needed information. OSHA can require employers to "make, keep and preserve" whatever records it may deem "necessary or appropriate for the enforcement of this Act or for developing information regarding the causes and prevention of occupational accidents and illnesses."[2] In addition, Congress <u>directed</u> the Secretary of Labor to "develop and maintain an effective program of collection, compilation, and analysis of occupational safety and health statistics" and authorized money, private research grants, and cooperative programs with the states in order to do so.[3]

OSHA has not properly utilized that authority. It's failure to carry out that mandate is probably the largest single weakness in the many efforts undertaken since 1971 to accomplish the Act's noteworthy purposes.

It virtually goes without saying that, to reduce the incidence of worker injuries and illnesses, you must first know their <u>cause</u>. With that information, you can

---

[2] Act, §(8)(c)(1), 29 U.S.C. §657(c)(1).
[3] Act §24, 29 U.S.C. §673.

determine (1) which OSHA standards are necessary and which are excess baggage, and (2) where changes or additions to the standards must be made. Knowing the causes of job injuries and illnesses will also permit OSHA to focus its enforcement resources upon the genuine problem areas. That kind of information would also permit employers to do likewise with their own safety and health programs.

Incredibly, however, OSHA either has not obtained statistics of that kind or has not utilized them for their obvious purposes. OSHA statistics can tell *how many* injuries and illnesses occur, in what *industries* they take place, and the *percent* of injuries and illnesses to total hours worked. But they don't identify what *caused* those injuries and illnesses. The shocking thing about this is that information on the causes of job injuries and illnesses *is compiled* and it *is* collected and analyzed. The states do it in connection with their worker compensation programs.

A separate Department of Labor agency, the Bureau of Labor Statistics (BLS), does utilize worker compensation data from the states to compile what BLS identifies as its Supplementary Data System (SDS). According to BLS, the SDS is "the *only* comprehensive source of information about factors associated with injuries and illness resulting from work-related accidents or exposures" and that it includes the "nature of injury or illness, part of body affected, characteristics of the accident or exposure which resulted in injury or illness" as well as information on the affected "industry, occupation, and characteristics of the affected workers.[4/] The "cause" of the injury or illness is not expressly mentioned in that description.

BLS concedes that SDS has its deficiencies because nearly one-third of the states did not participate in it at all and, that within the participating states, there are differences in reporting requirements. [5/] No one believed that it would be easy to develop the kind of statistical data that is essential to effective implementation of the Act, but it can be done, and Congress mandated that it would be done. Until it is, an accurate assessment of the Act's effectiveness in achieving its purposes cannot be made. And the absence of such data forces OSHA to virtually grope in the dark for a

---

[4/] Report of the President to the Congress on Occupational Safety and Health. 1980, Page 93 (President's 1980 OSHA Report). The edition of that Report covering calendar year 1987 makes the same statement although in somewhat different language. It also relates that 33 states participated in SDS in 1987. President's 1987 OSHA Report at 94.

[5/] President's 1980 OSHA Report at 94.

solution to the high worker injury/illness problem in this country.

Many examples to illustrate that point could be offered, one of which is the effectiveness of OSHA inspections. At its inception, OSHA identified five "target injuries" where injury rates were highest and concentrated its inspection activities there. It even supplemented its own inspectors by enlisting the inspection forces of the various states. When the program was completed, the injury rate in the five target industries showed an average 3.6% decline. The best showing was in lumber and wood products where the rate fell from 25.4 in 1972 to 24.1 in 1973.

During the same period, however, some industries _not_ targeted for OSHA inspections under the program showed an even _greater_ decline. For example, the transportation equipment industry rate fell from 18.8 to 16.7. See President's 1974 OSHA Report, pages 49-50. Twelve years after the discontinuance of the Target Industries Program, the injury rate for the lumber and wood products industry showed an even greater decline. In 1985, it was 18.2. President's 1987 OSHA Report at page 112.

Many people seem to assume that OSHA inspections are helpful in achieving the Act's goals. If that assumption was true, those businesses that have been most heavily inspected presumably would show lower injury/illness rates than comparable businesses that have not been so heavily subject to OSHA inspection. That comparison, however, cannot be made with the statistcal data available. [6/]

Despite the apparent lack of focus in the OSHA recordkeeping program, the Act imposes substantial recordkeeping requirements upon employers and OSHA has adopted implementing regulations that it enforces with inspections and citations. It is this matter to which we will now turn.

Employer Recordkeeping and Reporting Requirements

The Act directs OSHA to prescribe regulations requiring employers to maintain accurate records of work-related deaths, injuries and illnesses that involve loss of consciousness, restrictions of work or motion, transfer to another job, or medical treatment, except where such treatment is for minor injuries requiring only first aid.[7/] OSHA's implementing regulations appear in

---

[6/] The matters focused upon *during* OSHA inspections is also a subject that merits more attention. For example, during one calendar year OSHA issued 8,000 citations for noncompliance with an OSHA fan-guarding standard. There is no data, however, from which it can be determined how many, if any, employee injuries resulted from noncompliance with that standard - either in the year it produced 8,000 citations - or in any prior year.

[7/] Act, §8(c)(2), 29 U.S.C. §657(c)(2).

29 C.F.R. Part 1904. Both OSHA and BLS have also published explanations and guidelines in an effort to make those regulations more understandable. [8]

Both the relevant statutory requirement and OSHA's implementing regulations are phrased so generally that additional guidance is essential if uniformity is ever to be achieved and employers are ever to know what should and should not be recorded. Unfortunately, the "guidelines" issued to date [9] have not lived up to their purpose and may well have contributed both to employer confusion and an OSHA data base of questionable accuracy.

8/ They have **not** been adopted in accordance with the Secretary's rulemaking authority, however, so they may be changed from time to time without public notice and are of questionable validity. The OSH Review Commission has consistently held that guidelines for implementing the Act "do not have the force and effect of law." Secretary v. FMC Corporation, 5 BNA OSHC 1707, 1710 (1977). Pronouncements by the Secretary "that have not been promulgated as rules or regulations have no binding legal effect on either the Secretary or the Commission." Secretary v. Bristol Meyers Co., 7 BNA OSHC 1039 (n.1)(1978), emphasis added.

9/ In 1972, a booklet was published by BLS entitled, "What Every Employer Needs to Know About OSHA Recordkeeping" (Report 412). It claimed that it provided answers to questions most frequently asked by employers about recordkeeping and reporting of occupational injuries and illnesses. It was revised in 1973, 1975 and 1978. In 1986, it was replaced by another BLS booklet entitled, "A Brief Guide to Recordkeeping Requirements under the Williams-Steiger Occupational Safety and Health Act of 1970," revisions of which were published in 1975 and 1978 (the latter revision dropped the "Williams-Steiger" from the title). There is also a BLS booklet published in June 1983 with the same name as the 1978 OSHA booklet. In addition, the OSHA form 200 includes explanatory material on the recordkeeping requirements.

The recordkeeping regulations themselves require non-exempt employers[10] to maintain a log and summary of all "recordable occupational injuries and illness," 29 C.F.R. §1904.2, and to post in the workplace, from February 1 to March 1 each year, an annual summary of the included entries so employees can be informed. 29 C.F.R. §1904.5. In addition, employers must maintain a supplementary record of such mishaps. 29 C.F.R. §1904.4. This latter record covers the same kind of information normally recorded on worker's compensation and insurance reports. Indeed, those reports can be used to satisfy the §1904.4 requirements.

---

[10] Employers with ten or fewer full or part time employees are generally exempted from the recordkeeping requirement, as are employers engaged in retail trade, finance, insurance and real estate. 29 C.F.R. §1904.15(a) and 16. Establishments classified in Standard Industrial Classification Codes (SIC) 52-89 are also exempted, with the exception of those in SIC 52 (building materials and garden supplies), 53 and 54 (general merchandise and food stores), 70 (hotels and other lodging establishments), 75 and 76 (repair services), 79 (amusement and recreation services) and 80 (health services). However, those employers are obligated to report to OSHA any incident resulting in the death of an employee or the hospitalization of five or more employees. 29 C.F.R. § 1904.16. Those employers also may be obligated to keep records if they are selected to serve as the subject of a Bureau of Labor Statistics survey. 29 C.F.R. §1904.21. If any business or industry is so selected, it will be personally advised by BLS.

Some accidents must be reported to OSHA. That requirement is limited to any accident which results in either an employee fatality or in the hospitalization of five or more employees. 29 C.F.R. §1904.8. The report must be made within forty eight hours and it must detail the nature of the accident and the extent of the injuries or illnesses suffered therein.

Failure to keep required records or make such reports may result in the issuance of a citation to the offending employer. Indeed, multi-million dollar penalties have been proposed for the alleged failure to do so. See n.14 infra.

Both the Act itself and the implementing regulations authorize criminal penalties for employers who knowingly falsify OSHA records or reports. A maximum $10,000 fine and a 6-month jail sentence can be imposed. 29 U.S.C. 666(g), 29 C.F.R. §1904.9.

Employer Recordkeeping Guidelines

In fulfilling the employer's obligation to maintain a log and summary of all "recordable occupational injuries and illnesses," OSHA recommends that employers utilize OSHA Form No. 200 and, when preparing the supplementary information, the OSHA Form No. 101. However, any document that provides the same information in an equally comprehensible manner will suffice including worker compensation reports and insurance reports. Employers are obligated to make log and summary entries on form no. 200 within six working days of receiving notice of the injury or illness in question. 29 C.F.R. §1904.2(b)(1). A similar time limit applies for the completion of the OSHA 101 supplementary record forms. 29 CFR §1904.4.

Employers generally have not complained about the recordkeeping requirements per se. But they have complained that the guidelines that have been developed to assist them in the recordkeeping process to date are confusing, vague and ambiguous.

Many employers make the mistake of trying to find out what should be recorded by consulting only the booklets mentioned above (n.9). Their inquiry should <u>begin</u>, however, by consulting the regulations themselves. They do include definitions and they do not suffer from the legal infirmities noted above (n.8).

Only "recordable" injuries and illnesses have to be entered on the OSHA 200 or its equivalent. That term is defined as a <u>occupational</u> injury or illness <u>which results in</u> one or more of the following:

1. A fatality
2. One or more lost workdays
3. Transfer to another job
4. Termination of employment

5. Medical treatment other than first aid
6. Loss of consciousness
7. Restrictions on performance of normal job functions
8. Restriction of motion
9. A diagnosed occupational illness which is reported to the employer.

29 C.F.R. §1904.12(c). Three of those terms are further defined:

"Medical Treatment" - includes treatment administered by a physician or by registered professional personnel under the standing orders of a physician. Medical treatment does not include first aid treatment even though provided by a physician or registered professional personnel.

"First Aid" - is any one-time treatment, and any follow up visit for the purpose of observation of minor scratches, cuts, burns, splinters, and so forth, which do not ordinarily require medical care. Such one-time treatment, and follow up visit for the purpose of observation, is considered first aid even though provided by a physician or registered professional personnel. §1904.12(e).

"Lost Workdays"- The number of days (consecutive or not) after, but not including, the day of injury or illness during which the employee would have worked but could not do so; that is, could not perform all or any part of his normal assignment during all or any part of the workday or shift, because of the occupational injury or illness. §1904.12(f).

Those definitions are helpful and will provide an employer with the necessary guidance in obvious cases, such as an employee who breaks his leg in a fall from an elevated work station and cannot thereafter perform his normal job functions for several days or weeks. They leave literally thousands of cases in limbo, however.

It would not seem to be too difficult a task for OSHA to define more of its terminology -- such as "restriction of motion" - and to be much more specific in the definitions it does use, such as "medical treatment."

Interpreting the Recordkeeping Rules

The Act itself provides that, to be recordable, an injury or illness must be "work-related." One BLS publication states that work relationship is established under the OSHA recordkeeping system when the injury or illness results from an event or exposure in the work environment. The work environment is primarily composed of the employer's premises, and other locations where employees are engaged in work-related activities or are present as a condition of their employment. A work relationship is presumed when an employee is on the premises of the employer. That presumption may be rebutted by evidence that symptoms are exhibited which merely surfaced while the individual was on the premises and were not caused by any on-site experience. Work relationship must be proved when the employee is off premises.[11]

---

[11] "Recordkeeping Requirements for Firms Selected to Participate in the Annual Survey of Occupational Injuries and Illnesses, Covering 1987." O.M.B. No. 1220-0029, p.6, 7.

The guidelines also recommend that employers engage in a five step analysis when deciding whether an injury or illness is recordable or not. Employers should initially determine whether in fact a recordable instance has occurred at all, that is, whether there was a death, an illness, or an injury. If the answer is "yes" the employer must then decide if the case occurred in the work environment. Upon a determination that it has, the employer must then determine if the case is an injury or an illness. If the employer concludes that the incident was an illness, he then must record it in the portion of the OSHA Form No. 200 devoted to illness reports. If the employer decides that the incident was an injury, he must record it in the injury portion of the OSHA Form No. 200 if he determined that the injury is "recordable."[12/]

12/ Id. at 5. It may be of interest to note that those guidelines regularly use such terminology as: "If the employer determines, etc." All of the guidelines published to date either state or infer that the judgement as to whether or not a particular incident is "recordable" is for the employer to make. Thus, it would seem that an employer's honest mistake of judgement should not be the subject of an OSHA citation. That has not, however, been the case.

Medical Treatment/First Aid

Distinguishing between medical treatment and first aid can cause the employer a headache that itself may necessitate "medical treatment."[13] Generally speaking, one-time visits to a medical facility or subsequent vists requiring mere observation by medical personnel are not considered to constitute "medical treatment." Subsequent visits for further treatment do constitute "medical treatment" and must be recorded. However, the boundary between what constitutes first aid and what constitutes medical treatment is often a rather tenuous one. Distinguishing between recordable and non-recordable injuries solely on the basis of the treatment provided remains one of the many imponderables in OSHA recordkeeping.

13/ The 1986 BLS publication listed in n.11, supra, states that medical treatment is involved when the injury or illness requires the use of any prescription medicine except a single dose administered on a first visit for minor injury or discomfort. Thus, if an injured employee goes to a medical facility and is told he can return to work and is also given a prescription to take if his pain continues overnight, there is no "medical treatment until the employee gets that prescription filled and takes that medicine. Because circumstances such as this are not unusual, it appears that a detailed inquiry must be made of the employee. Requiring each employee who reports an injury or illness to fill out a questionnaire might be helpful. A suggested questionaire is printed in Appendix I, Section A, of OSHA Handbook, Second Edition, published by Government Institutes, Inc., in 1989.

Consider the following situation. An employee receives a cut on his hand in the course of the performance of his duties. He visits a doctor. If the visit results in the application of an antiseptic to the wound, the treatment provided would be considered first aid rather than medical treatment - therefore nonrecordable. If there is then a follow-up visit by the employee at which the doctor or nurse merely observes the cut without reapplying antiseptic, the injury would still be considered nonrecordable. However, if the doctor or nurse reapplies antiseptic to the cut as a precautionary measure during the second vist, OSHA would regard this procedure as the provision of medical treatment and thus a recordable injury. The employer would therefore be obliged to enter the cut hand incident on his OSHA form 200.

Restriction of Motion

Whether there exists a work-related "restriction of motion" is another imponderable that has produced considerable confusion, particularly among employers making good faith efforts to safeguard their employees' health. Employers in a variety of industries where employees regularly make use of a particular set of muscles have instituted job rotation programs to prevent employee fatigue and motion trauma. That process can be analogized to a situtation most people have experienced when carrying a heavy suitcase: they continually switch the suitcase from one hand to another as an arm becomes tired.

Employers institute job rotation programs to <u>avoid</u> employee injury. However, some OSHA people consider rotation programs as <u>proof of injury</u>. Under that view, the reason for switching the heavy suitcase from the right hand to the left hand is because of "restriction of motion" of the right hand. Employers have been cited for not recording matters like that on the OSHA 200 form.

Examples such as these exemplify employer exasperation with the recordkeeping requirements. Most employers make a good faith effort to safeguard the well-being of their workers and are willing to adhere to clear requirements that serve that end. However, instances of uncertainty in determining OSHA "recordability" are a frequent occurrence. Ill-defined OSHA statements on what constitutes a recordable injury or illness threaten to transform employer-maintained OSHA 200 forms into useless, repetitive, multivolume compilations, thereby undermining their ability to contribute towards the worthwhile goal of precisely defining the true occupational safety and health problem as mandated by Congress.[14]

---

[14] During the first 15 years of its existence, numerous instances of employer failure-to-record were disclosed during OSHA inspections. Citations were issued. Some contained no proposed penalties - others had nominal penalties. In 1985, however, OSHA made a profound change in that practice. It was first manifested by a $1.37 million penalty proposal against the Union Carbide Corporation, a substantial proportion of which was based upon alleged recordkeeping violations. No public announcements of the reason for this sudden turnabout was made. It obviously resulted from a policy change, however, because numerous other companies have since been the recipient of substantially similar OSHA citations. For example, the August 3, 1987 edition of the Washington Post listed the "Top 10 OSHA Fines Imposed since April, 1986." They ranged from a low of $463,500 to a high of $2.6 million. Seven of the ten were listed as recordkeeping violations. During the last three years (1987, 1988 and 1989), OSHA proposed $12 million in penalties for alleged recordkeeping violations.

The practice of seeking unusually stiff penalties for alleged recordkeeping noncompliance further exacerbates the problem of developing accurate information on work injuries and illnesses because it forces employers to either record all injuries and illnesses regardless of cause or work-relatedness, or to resort to double bookkeeping. It is quite unlikely that any improvement in injury/illness records will result from the current OSHA practice. Indeed, for reasons discussed later, the result may be just the opposite.

Recordkeeping Compliance

The change in OSHA's recordkeeping citation policy noted above (n.14) provoked a groundswell of industry criticism charging OSHA with citation overkill. There also was unaccustomed attention to OSHA matters in the Congress, the public media and other quarters. Understandably, companies do not want to receive such citations but -- in view of the indefiniteness in the recordkeeping regulations -- there can be no assurance that similar citations will not issue against them. Of course, they can contest such citations and put OSHA to its proof. That will make it very difficult for OSHA and very expensive for both OSHA, the cited employer, the OSH Review Commission and the court system.[15/]

In an attempt to avoid the kind of citation and penalty referenced above (n.14), some companies have indicated that they will henceforth resort to a dual recordkeeping system under which those cases thought to be "recordable" by OSHA would be entered on the OSHA 200 Form while a separate system would be maintained for the company's own use to record only the "real" injury/illness cases that are both occupational and work-related.[16/]

15/ OSHA's problems in proving a contested recordkeeping citation can be demonstrated, at least partially, by the decision in Brock v. Dow Chemical, U.S.A., 801 F.2d 926 (7th Cir. 1986).

16/ It should be noted that concerned employers have maintained injury/illness records long before there was any OSHA requirement to do so. See the discussion in Secretary v. Taft Broadcasting Company, Kings Island Division, 13 BNA OSHC 1137, 1140-1141 (1987). They are quite useful in identifying and correcting hazards within the company and thus are helpful to both the employer and its employees. In addition to the obvious benefit of a less hazardous workplace and fewer good employees lost, lower worker compensation costs can be realized.

The reasoning behind this move to recordkeeping duality goes something like this: To date, there has been no company penalized for overrecording. Thus, record everything imaginable on the OSHA 200. That will insulate the company from OSHA recordkeeping penalties. It is regretable that such a practive will also destroy the very reason why a recordkeeping requirement was included in the Act, but why risk millions in penalties by trying to do it right when there is nothing to lose by overrecording?

Industry criticism of OSHA's recordkeeping citation practices has been met with public statements from OSHA, the media, and a number of others that the cited companies are intentionally "misrepresenting" their actual injury/illness experience.[17] Some of those commentators have suggested that companies have done that in order to avoid an OSHA inspection.

That rationale will not withstand analysis in most of industry because the cut-off point for avoiding an OSHA inspection -- which in 1987 was a lost workday injury

---

17/ See, for example, the quoted statement of former Assistant Labor Secretary John A. Pendergrass, USDL News Release 87-167, April 24, 1987, commenting upon OSHA recordkeeping citations issued to John Morrell & Company.

(LWDI) rate of 3.4[18]/ -- is far below the 15.1 LWDI rate in manufacturing, construction, maritime and similar sectors of "industry."[19]/ Few, if any, companies that have received high-penalty recordkeeping citations recently have a 3.4 rate or below. Clearly, there is no incentive for such a company to underreport in order to achieve a 7.5 rate, for example, nor any disincentive for reporting a 28.7 or higher rate if that is indeed their actual experience. So long as the rate reported is 3.4 or higher, the company is regarded as "high hazard" under the

---

18/ See OSHA Instruction CPL 2.25G, Oct. 1, 1986, "Scheduling System for Programmed Inspections," Appendix C. Use of injury rate tabulations as an OSHA inspection-targeting device was adopted in 1981 by then Assistant Secretary Thorne Auchter. It was the subject of considerable criticism from organized labor and, at the present time, is not uniformily observed.

19/ The word "industry" for OSHA statistical purposes includes white-collar services like real estate, finance, accounting, law firms and employers engaged in similar enterprises where injury/illness rates are virtually nonexistent. By including such firms in its calculation of any "average" for all "industry" the incidence rate becomes considerably lower than that of most companies employing "blue collar" workers.

OSHA inspection targeting program.[20] Claims of underreporting in order to avoid OSHA inspection, therefore, are simply empty rhetoric if the company's reported rate was 3.4 or above.

Even a reported rate lower than 3.4 will not prevent an OSHA inspection. Use of a cut-off rate applies only to safety - not occupational health - inspections and has no application at all where an inspection results from an employee complaint, a referral, an imminent danger, a fatality, or catastrophe. It doesn't even apply to all routine "safety-only" inspections. See the documents referenced above at n.18 and n.20. Moreover, the use of a cut-off point for routine OSHA safety inspections is a matter of policy - not a requirement of law. It was first adopted in 1981 in order for OSHA "to maximize the utilization of available resources." OSHA Instruction PAE 1.1A, Sept. 20, 1982. See also n. 18, supra. That policy could be changed by OSHA at any time it chooses. Presumably, that would be done if it were being abused by business establishments.

---

[20] OSHA Instruction CPL 2.45A, March 27, 1986, Chapter II, "Compliance Programming."

There is another plausible argument to be made in favor of <u>overreporting</u> in order to avoid astronomical OSHA penalties even if it would produce an inspection that might otherwise have been avoided: The highest reported OSHA penalty for workplace <u>hazards</u> that could cause death, injuries and illnesses is more than one million dollars <u>less</u> than the highest <u>recordkeeping</u> penalty. <u>See</u> n.14, <u>supra</u>.

One of the declared Congressional purposes for achieving the noteworthy objectives of the Occupational Safety and Health Act was to put in place a procedure that would "<u>accurately</u> describe the nature of the occupational safety and health problem." 29 U.S.C. §651(12), emphasis added. Double bookkeeping, underreporting and overreporting will undermine those objectives. To know whether either of the latter has occurred, however, it is necessary to first know what constitutes "accurate" reporting. It would seem to be clear from the discussion above that the criterion presently used for making that determination is much too indefinite to achieve the recordkeeping accuracy that Congress felt was essential.

Understanding the Rules

OSHA recordkeeping regulations and the guidelines used to explain them are not only indecipherable to employers but, in OSHA itself, it is difficult to find any two people who will always agree on what should be recorded, and if there should be agreement on recordability, reasonable OSHA inspectors may differ upon where in particular such an incident should be entered on the OSHA 200 Form.[21]

---

[21] Some recordkeeping citations allege failure to record an incident in the "right place" on the form, not failure to record. It is not always easy to distinguish between an injury and an illness but they must be listed at different places on the OSHA 200 form. Here is an example: There is a current OSHA emphasis program on cumulative trauma disorders (CTD). OSHA claims that a CTD develops as a result of chronic exposure of a particular body part to repeated biomechanical stress, which by cumulative effect, produce a debilitating physical condition. Some examples are: carpal tunnel syndrome, tendinitis and low back pain. Question: Is CTD an injury or an illness? OSHA says that CTD conditions constitute "occupational illness" for OSHA-200 purposes and must be entered in the "7f" column on that form -- but back injuries that develop as a result of repeated trauma are injuries and must be recorded as such.

The June 1986 Labor Department booklet issued as a Guide to the OSHA Recordkeeping Requirements (n.11, supra) for example, states that employee hospitalization for observation "is not recordable, provided no medical treatment was given, or no illness was recognized." Yet there are a number of reported cases where employers have been cited on just such facts.[22] Obviously there are a number of people within OSHA itself who interpret that requirement differently. Employers are also confused on this point, and as a result, the recordkeeping purposes are not being properly served.

As mentioned earlier, OSHA defines a recordable case in a number of different ways including "restriction of work or motion" and "transfer to another job" even though no lost work time is involved. 29 C.F.R. §1904.12(c)(3). The failure to be more specific in this regard creates real problems, especially in labor-intensive industries.

On the recommendation of ergonomic authorities, many companies have installed job rotation programs in order to

---

[22] Secretary v. Western Airlines, Inc., 12 BNA OSHC 1084 (1984), Secretary v. General motors Corp., 11 BNA OSHC 2013 (1984), and Secretary v. Simplex Time Recorder Co., 11 BNA OSHC 1758, 1763 (1983), aff'd sub nom Simplex Time Recorder Co. v. Secretary of Labor, 766 F. 2d 575, 590 (D.C. Cir. 1985).

avoid motion trauma. Thus, an employee will, from time to time, be switched to job assignments that utilize different muscles. That may occur several times a day for a large number of employees. As mentioned earlier, however, there are OSHA people who claim that the practice itself falls within the definition of recordable injury just referenced (§1904.12(c)(3)). If that interpretation is correct, then the OSHA 200-Form would cover volumes rather than pages and may not constitute an accurate report of injuries. Nevertheless, some companies are apparently recording these job changes and some are not. An accurate description of the occupational safety and health problem is impossible when there is a confusing and ambiguous situation like this one.

Hundreds - if not thousands - of similar examples of uncertainties in the recordkeeping requirements could be cited. It is doubtful whether a recordkeeping citation based upon situations such as those would satisfy the rule that employers - and all other citizens of this country - are entitled to fair warning of exactly what the law requires or prohibits.[23]

---

[23] See, for example, Kropp Forge Company v. Secretary of Labor, 657 F.2d 119, 122 (7th Cir, 1981).

The vast majority of employers observe the OSHA recordkeeping requirements as best as they know how. They should have no qualms about increasing the number of cases listed on the OSHA 200 forms if that is what OSHA desires. If an employer decided to record every hint or suggestion of employee injury or illness on the OSHA 200 form, his incidence rate would skyrocket but that would not produce any adverse consequences to the employer so long as those who set worker compensation rates understood that the rate does not show the "real" job injury experience but results simply from the employer's defensive strategy against million dollar OSHA recordkeeping citations. [24/] The real victim of that strategy - and the only victim - is the purpose sought to be achieved by the OSH Act itself.

Another employer compliance strategy in response to OSHA's post-1985 recordkeeping citation practices involves more - and different - records than employer's have

---

24/ An employer who follows that method but also wants it known that he does not think the incident is recordable can write that right on the OSHA 200 form in any manner or with any words of his choice. So long as the information the form seeks is recorded, the legal requirement is satisfied. Any additional information, comment, protest or opinion will not change that. OSHA people have suggested that questionable incidents should be first entered on the OSHA-200, then red-lined.

heretofore maintained. Some of them employ detailed questionnaires that are executed by the injured employee, his supervisor and the treating nurse or physician (if any). If <u>any</u> of the information obtained by that procedure indicates a recordable injury or illness, it is entered on the OSHA 200 Form. A copy of a questionnaire of that kind was referenced <u>supra</u> at n.13.

<u>Where to Keep OSHA Recordkeeping Forms</u>

Section 1904.2 requires employers to maintain "in each establishment" the previously described log and summary of employee injuries or illnesses (OSHA forms 200 and 101) "for that establishment."[25] However, because many covered employees venture far afield from their employer's headquarters in the course of their employment, §1904 adopts a two-pronged definition of "establishment" that distinguishes between employers who operate from a fixed location and those who do not. Employers who operate from a fixed location should keep their OSHA no. 101 and 200

---

[25] In OSHA lexicon the word "establishment" is a word of art. Those who don't realize that fact often confuse it with terms such as "business" or "industry". It is important, therefore, to recognize and understand the differences. For OSHA purposes the word "establishment" means "<u>a single physical location</u> where business is conducted or where services or industrial operations are performed." 29 C.F.R. §1904.12(g)(1), emphasis added.

forms there. Records for employees who report to the fixed location before venturing elsewhere should also be kept at the fixed location. The records of employees who do not operate from any single fixed location should be kept at the site where they are paid or where their other personnel records are stored.

Section 1904.14 specifies where the forms are to be maintained for employees who are not associated with a particular fixed location but are subject to common supervision. Employers may store the records of those employees at a central information storage location as long as the address and telephone number of that central location is available at each work site where employees are engaged and the central location is manned during normal business hours.

Records Retention

Employers are obligated to maintain their OSHA 200 and 101 forms for a period of five years beyond the date to which they relate. 29 CFR §1904.6. An employer acquiring the interest of a previous employer's records is obligated to retain them for a similar period but is not obligated to update such records. §1904.11. The length of time other OSHA-required records must be retained varies from standard to standard. The retention period is 30 years for employee "exposure" and "medical" records. The regulations covering those records are discussed infra.

OSHA Inspection of Injury/Illness Records

In the Taft Broadcasting case, supra, 13 BNA OSHC 1137, the OSH Review Commission ruled that OSHA has no right to inspect an employer's injury/illness records unless the employer voluntarily discloses them to OSHA, or an OSHA inspection warrant or subpeona is issued that requires the employer to permit such a records inspection.

OSHA appealed that decision but the Court of Appeals refused to overturn the Commission's ruling. It held, rather, that "employers have a recognizable privacy interest in the records in question even though the employer is required by law to keep them." McLaughlin v. Kings Island, Div. of Taft Broadcasting Co., 849 F.2d 990, 995 (6th Cir. 1988). The Court decision went on to point out that, when OSHA wants to inspect the OSHA 200 Form or similar records:

> "The Fourth Amendment requires that the employer have some notice and opportunity to be heard to challenge the resonableness of the agency request. While an 'agency has the right to conduct all reasonable inspections of such documents which

> are contemplated by statute...<u>it must delimit the confines of a search by designating the needed</u> documents in a formal subpeona."

<u>Id.</u>, at 996, quoting from <u>See v. City of Seattle</u>, 387 U.S. 541, 544-545 (1967), emphasis in the original.

The Court agreed with the Commission's ruling that "insofar as 29 C.F.R. §1904.7(a) authorizes the Secretary to obtain access to business records without either a warrant or an administrative subpoena...that it is inconsistent with the Fourth Amendment of the United States Constitution." <u>Id.</u>, 849 F. 2d at 997.

There had been a prior opinion contrary to <u>Kings Island</u> by a different appellate court, <u>McLaughlin v. A.B. Chance Co.</u>, 842 F. 2d 724 (4th Cir. 1988). The Sixth Circuit acknowleged that decision but declined to follow it by noting that it made no mention of a "controlling" Supreme Court decision: <u>New York v. Burger</u>, 107 S. Ct. 2636 (1987). A different appellate court sitting in Atlanta has also agreed with the Commission on the issue. <u>Brock v. Emerson Electric Co.</u>, 834 F. 2d 994 (11th Cir. 1987). The <u>Kings Island</u> court cited that decision with approval.

As a result of the foregoing, there has been an increasing trend among employers since mid-1988 to "just say no" when an OSHA Compliance Officer says he wants to look over their OSHA 200 Forms and other injury/illness records but does not have a warrant or subpoena authorizing a records inspection.

Access to Records

The OSHA injury/illness forms discussed above are, of course, not the only records that employers keep in the regular course of business. Many have records that under ordinary circumstances are confidential, such as employee physical exams, or other medical information on employees, job applications, records of salary and wages, performance evaluations, etc. Increasingly, however, attempts are made by government, labor unions and others to obtain those records as well as other records in the employer's possession. The subject is very sensitive to many employers and provokes considerable controversy and litigation.

Section 8(c)(3) of the OSH Act, 29 U.S.C. 657(c)(3), provides that the Secretary of Labor:

> "[S]hall issue regulations requiring employers to maintain accurate records of employee exposures to potentially toxic materials or harmful physical agents which are required to be monitored or measured under section 6.[26] Such regulations shall provide employees or their representative

---

[26] Section 6 is the provision authorizing the adoption of OSHA standards, so the use of the term "section 6" in this section should be read to mean "the OSHA standards."

with an opportunity to observe such monitoring or measuring, <u>and to have access to the records thereof</u>. Such regulations shall also make appropriate provision for each employee or former employee to have access to such records as will indicate his own exposure to toxic materials or harmful physical agents in concentrations or at levels which exceed those prescribed by an applicable occupational safety and health standard promulgated under section 6, and shall inform any employee who is being thus exposed of the corrective action being taken."

Emphasis added. The foregoing is rather narrowly drawn because it specifies only employee exposures to "toxic chemicals and harmful physical agents" which are "required to be monitored or measured" by an OSHA standard. <u>See</u> n. 33, <u>infra</u>. However, another provision of the Act authorizes the Secretary to "prescribe such rules and regulations as he may deem necessary to carry out his [or NIOSH's] responsibilities under this Act."[27]

---

[27] Act §8(g)(1), 29 U.S.C. §657(g)(1). The term "NIOSH" is the acronym for the National Institute for Occupational Safety and Health, an agency of the U.S. Department of Health and Human Services.

OSHA has used that authority to promulgate so-called "records access" regulations. They will now be discussed.

Who Has Access to Records?

Employees are generally recognized to have a right of access to at least some company records that pertain to them. §1904.7 specifies that employers must make their form no. 200 logs and summaries available to current and former employees and/or their representatives for their inspection. That provision expressly recognizes the right of employees to bargain collectively with employers to obtain access to employee records relating to occupational injury and illness. 29 CFR §1904.7(b)(2). Thus, collective bargaining agents may serve as "employee representatives". Employers are further obligated to provide form no. 200 logs and summaries as well as form no. 101 supplementary reports to designated governmental officials.[28]

Labor unions and so-called community or environmental "actvists" often are particularly anxious to obtain access to company records. During the years when the OSHA leadership was particularly sensitive to the wishes of those groups, the agency adopted a far-reaching regulation entitled "Access to employee exposure and medical records." It appears today , essentially unchanged from the form in which it was promulgated in 1980, as 29 C.F.R. §1910.20. Some of its far-reaching provisions will be pointed out *infra* but there is much more to it than can be covered here. Employers should carefully review the regulation itself and relevant judicial interpretations

---

28/ OSHA rules provide that employers are obligated to provide requested records for inspection and copying to any representative of the Secretary of Labor for the purpose of carrying out the provisions of the OSH Act, any representative of the Secretary of Health and Human Services conducting an investigation pursuant to section 20(b) of the OSH Act, or any representative of a state that has jurisdiction for occupational safety and health inspections or statistical compilations. However, there are constitutional, regulatory and judicial restrictions on this rule that should be considered by the employer before granting access to records. They were discussed above at pp. 31-33.

thereof (as well as the matters noted above at n.28) whenever a "records access" question comes up.

There are also specific records-access provisions included within the text of many OSHA standards. For example, the hearing conservation standard requires the employer to maintain records of noise exposure measurements and employee audiometric testing, 29 C.F.R. 1910.95(m), and then provides that: "All records required by this section shall be provided upon request to employees, former employees, representatives designated by the individual employee, and [OSHA inspectors]." Section 1910.95(m)(4).

Requirements of the Records-Access Regulation

Representatives of business and industry launched an immediate court challenge to §1910.20 when it was adopted in 1980 but it was unsuccessful.[29]

---

[29] See Louisiana Chemical Association v. Bingham, 550 F. Supp. 1136 (W.D. La. 1982). That decision was affirmed without further discussion by the Court of Appeals (Fifth Circuit) in 1984. The regulation was also at issue in an earlier decision under the same name. It appears in 657 F.2d 777 (5th Cir. 1981).

The records access regulation obligates employers to preserve and retain employee exposure[30] and medical[31] records, and to make them available to

___

[30] "Employee exposure records" is defined in §1910.20(c)(5) to include any record containing information on employee exposure to a "toxic substance or a harmful physical agent." OSHA interprests that to mean that any information derived from environmental monitoring or sampling as well as related data relevant to calculations based thereon are part of an employee exposure record and would be obtainable. The same goes for biological monitoring results assessing the absorption by the body of a substance or agent as measured by physical samples (as, for instance, a blood sample), the Material Safety Data Sheets required by OSHA's Hazard Communication Standard, and any other record that reveals the identity of a toxic substance or harmful physical agent.

[31] Employee Medical Records" is defined in §1910.20(c)(6) to include any record concerning the health of an employee made or maintained by health care personnel. OSHA interprets that to mean that medical and employment questionnaires or histories, medical exam and lab test results, medical opinions/diagnoses/ progress notes/recommendations, descriptions of treatments and prescriptions, and employee medical complaints are obtainable by employees or their representatives. §1910.20(c)(6)(ii) lists some things that are not included in the OSHA definition of that term. They include: physical samples that are not legally required and which are routinely discarded, records concerning health insurance claims that are maintained separately from the employer's health plan and which are non-accessible to the employee normally, and records concerning voluntary employee assistance programs maintained separately from the firm's medical program.

employees or their designated representative for their examination and copying. 29 CFR §1910.20(e). It also obligates covered employers to provide those records to OSHA representatives upon request. However, there are court cases (discussed above) that limit such obligations and there is another regulation, 29 C.F.R. §1913.10, that contains restraints upon OSHA access to employee medical records. That regulation should be consulted whenever disclosure of any such record is under consideration.[32]/ Employers should also bear in mind that improper release of records - even to government agents - may provoke civil litigation against them.

---

[32]/ The regulation includes the following statement that is indeed true (or should be) no matter who is seeking access to those records: Medical records "contain personal details concerning the lives of employees. Due to the substantial personal privacy interests involved, OSHA authority to gain access to personally identifiable employee medical information will be excercised only after the agency has made a careful determination of its need for this information, and only with appropriate safeguards to protect individual privacy." 29 C.F.R. §1913.10(a). Whenever an OSHA inspector seeks access to such records, he is required, among other things, to obtain the signed authorization of both the OSHA assistant secretary and a designated medical doctor and to provide them to the employer who has the records being sought. See n. 35, infra.

The purpose of the OSHA records-access rule is to facilitate the identification of both the scope of, and the adverse health impacts emanating from, employee exposure to "toxic substances and harmful physical agents."[33/] The regulation provides employees a "central role in the detection and solution of health problems as there are no assurances that anyone else will protect their health with equal vigor or determination." 45 Fed. Reg. 35213 (1980). Thus, the rule attempts to create an important discovery role for the most motivated party: the worker himself. However, the most activity to date under the regulation has resulted from labor unions.

The regulation binds current, former or successor employers of employees exposed to toxic substances or

---

33/ "Toxic substances and harmful physical agents" are defined as any chemical substance, biological agent (bacteria, virus, etc.), or physical stress (noise, heat, cold vibration, repetitive motion, ionizing and non-ionizing radiation, hypo- or hyperbaric pressure, etc.) that is regulated by Federal law or rule due to a hazard to health; is listed in the latest printed edition of the NIOSH Registry of Toxic Effects of Chemical Substances (RTECS); has yielded positive evidence of an acute or chronic health hazard in human, animal, or other biological testing conducted or known by the employer; or has a material safety data sheet available to the employer indicating that the material poses a health hazard. 29 CFR §1910.20(c)(11).

harmful physical agents "whether or not the records are related to specific [OSHA]] standards," 29 CFR §1910.20(b)(2), and obligates them to do the following:

    1. Inform their employees of their right of access to designated records when they enter the employer's service and remind them of their right of access at least annually thereafter.[34/] 1910.20(g).

    2. Maintain employee exposure records for a period of at least thirty years. 1910.20(d)(1)(ii).

    3. Maintain employee medial records both for the entire duration of the employee's term of employment and for an additional thirty year period after termination of employment. 1910.20(d)(1)(i).

    4. Provide employees and employee representatives with access to such records in a reasonable manner no later than 15 days after a request for such access has been made.[35/] 1910.20(e).

34/ It appears that very few employers are aware of that requirement because noncompliance is widespread. There is a suggested form and procedure included in Appendix I, Section B, of OSHA Handbook noted supra (n.13), that will fulfill that obligation as well as a number of similar obligations that are contained in various OSHA standards and regulations.

35/ If further provides, however, that employers must provide OSHA officials with immediate access to such records. That provision should not be taken literally. Neither an OSHA inspector nor a local Area Director is permitted access to the employee medical records referenced in 1913.10 without specific written authorization from OSHA's Assistant Secretary of Labor. That written authorization must be served upon the employer. Moreover, employers requested to produce records have Constitutional rights to refuse. See the discussion at pp. 31-33, supra.

Court Interpretation of Records-Access Rule

Some aspects of the Records Access Rule have sparked criticism from employers. They decry both the ambiguity and the all-inclusive nature of several of the provision's key terms as well as the administrative costs and burdens imposed by the Rule. Some fear that union and employee access to such a wide range of company records inevitably will result in the disclosure of company trade secrets.

Other employers feel that the disclosure of potentially sensitive employee medical records to government investigators and unions violates the privacy rights of employees and actually creates a disincentive for employees to participate in employer-maintained health programs. It is also feared that such disclosures invite the filing of harassment-based individual and class action lawsuits. There have been court decisions on a number of those concerns.

It was held in U.S. v. Westinghouse Electric Corporation, 638 F.2d 570 (3rd Cir. 1980), that an employer could not unilaterally withhold employee medical records requested by NIOSH simply by invoking, on their behalf, the employees' right to privacy in the records.

The court agreed with Westinghouse, however, that all medical records contain information that must in some sense be regarded as sensitive, but noted that individuals possess no <u>absolute</u> right to privacy in their medical histories and that the right to privacy may be outweighed by a compelling public interest favoring disclosure.

The court found that the requested records at issue (results of routine blood tests and cardiograms) were <u>not</u> the type of sensitive personal data about which employees were likely to be concerned, but it did reserve the right of individual employees to themselves invoke their privacy right. Thus, empoyers who are asked for access to records containing personally identifiable employee medical data should have the person making such a request produce a waiver signed by each subject employee. The <u>Westinghouse</u> decision suggests that precaution.

Whether or not an employer must produce requested records will frequently be determined by the nature of the records, what they contain, and whether the records are kept for the employer's information or because he is required by law to do so. In <u>Establishment Inspection of Kulp Foundry</u>, 691 F.2d 1125 (3d Cir. 1982), the Court held that an employer did not have to give OSHA requested records even though the inspection was conducted pursuant to a warrant that expressly authorized OSHA to do so. It held that OSHA must use the subpoena process to obtain such records.

A later decision, Donovan v. Wollaston Alloys, Inc., 695 F. 2d 1 (1st Cir. 1983), distinguished between records required to be maintained by OSHA regulation (such as the OSHA 200 form) and other company records (such as crane maintenance records), and held that the company had to allow OSHA access to the former but not the latter. However, in Taft Broadcasting, and the related cases discussed above, which constitute the most recent court rulings on the matter, it was held that the OSHA 200 form was not a record maintained exclusively for OSHA purposes and thus OSHA could not obtain access to it without the employer's consent unless it first obtained court authorization.

Requests for records made by a labor union, whether or not the records are required to be maintained only for OSHA purposes, may be subject to the provisions of the National Labor Relations Act and decisions issued pursuant thereto. Such requests should be considered and acted upon accordingly.

In concluding this discussion, there is one rather important point to be made. Records hold a very special place in the eyes of the law. Employers should not provide access to them willy-nilly even when the request comes from the government itself.

RECORDKEEPING REQUIREMENTS UNDER SARA TITLE III.

J. Gordon Arbuckle, Timothy A. Vanderver, Jr., Paul A. J. Wilson*

The major recordkeeping provisions of Sara Title III are found in subtitle B, which contains three distinct provisions of this type.[1]/ The first two of these are basically complementary. Section 311 requires facilities at which "hazardous chemicals" are present in excess of specified thresholds to submit to the LEPC, SERC, and local fire department copies of the material safety data sheets ("MSDS"), or a list of the substances (grouped into hazard categories) for which they maintain MSDSs. This reporting requirement gives emergency planners and response entities notice of the types of hazards presented by materials at a facility.

Section 312 provides emergency planners with additional information on hazardous chemicals by requiring the facility owners and operators required to submit an MSDS or a list of MSDS chemicals under Section 311 to file an annual inventory form with the LEPC, SERC, and local fire department. This inventory provides an estimate of the maximum amount of hazardous chemicals present at the facility during the preceding year, an estimate of the average daily amount of hazardous chemicals at the facility,

---

[*]/ Patton, Boggs & Blow, Washington, D.C.

[1]/ Subtitle A principally concerns emergency release reporting for so-called "extremely hazardous substances," and Subtitle C contains miscellaneous provisions concerning, for example, trade secrecy, penalties and causes of action for non-compliance, and providing information to health-care

Continued

and the location of these chemicals at the facility. Inventory information is reportable in two forms. "Tier I" reports contain general information on the amount and location of hazardous chemicals by category and are submitted annually. "Tier II" reports provide more detailed information on individual hazardous chemicals, must be submitted by the owner or operator upon request, and may be submitted in lieu of the Tier I report.

Section 313, the third reporting provision of Subtitle B, is significantly different from Sections 311 and 312. This provision requires covered facilities to report annually to EPA and the state in which the facility is located on all environmental releases of specified "toxic chemicals" that are manufactured, processed, or otherwise used at the facility in excess of certain threshold quantities. These reports make available to regulatory authorities, as well as to any interested citizen or citizens group, hitherto unavailable data on the identity and quantity of a large number of chemicals that are released into the environment by manufacturing sector facilities.

A. Section 311 -- Material Data Safety Sheets

Subject to quantity-based reporting thresholds set by EPA, the owner or operator of a facility who is required by OSHA's hazard communication regulations[2]/ to generate or maintain an

---

professionals.

[2]/ 29 C.F.R. Parts 1910, 1915, 1917, 1918, 1926, and 1928, 52 Fed. Reg. 31852 (August 24, 1987). These standards
Continued

MSDS[3]/ for any "hazardous chemical" must provide either a copy of each MSDS or a list of the chemicals for which an MSDS is prepared or maintained to the LEPC, the SERC, and the local fire department with jurisdiction over the facility. Section 329(5) of EPCRA defines "hazardous chemical" as having the same meaning given the term by Section 311(e) of the Act. Section 311(e), in turn, defines the term as having the meaning given it by the OSHA hazard communication regulations, with certain specified exceptions. The OSHA rule defines the term as "any chemical which is a physical hazard or health hazard."[4]/ A "physical hazard" is "a chemical for which there is scientifically valid evidence that it is a combustible liquid, a compressed gas,

---

    originally required chemical manufacturers and importers to obtain or develop an MSDS for each hazardous chemical they produced or imported. Additionally, all manufacturing sector employers in SIC Codes 20 through 39 were to provide, *inter alia*, information to their employees about the hazardous chemicals to which they were exposed through the use of MSDSs obtained from the chemical manufacturers or importers from whom they purchased the subject hazardous chemicals. Distributors of such substances were required to supply copies of MSDSs they received to other distributors and manufacturing sector purchasers. However, in the August 24, 1987, rulemaking, OSHA expanded the scope of the hazard communication rules to cover all employers with employees exposed to hazardous chemicals in their workplaces. As of January 30, 1989, following a series of challenges, these regulations applied to all employers.

[3]/   § 329(6), 42 U.S.C. § 11049(6), defines the term "material safety data sheet" to have the same definition as that promulgated by OSHA in its hazard communication rules. Those rules define the term to mean "written or printed material concerning a hazardous chemical which is prepared in accordance with Paragraph (g) of this section." *See, e.g.*, 29 C.F.R. § 1910.1200(c).

[4]/   29 C.F.R. § 1910.1200(c).

explosive, flammable, an organic peroxide, an oxidizer, pyrophoric, unstable (reactive) or water-reactive."[5] A chemical that is a "health hazard" is one:

> for which there is statistically significant evidence based on at least one study conducted in accordance with established scientific principles that acute or chronic health effects may occur in exposed employees. The term "health hazard" includes chemicals which are carcinogens, toxic or highly toxic agents, reproductive toxins, irritants, corrosives, sensitizers, hepatotoxins, nephrotoxins, neurotoxins, agents which act on the hematopoietic system, and agents which damage the lungs, skin, eyes, or mucous membranes....[6]

It is very important to note, however, that a wide range of materials are excluded from this definition. Section 1910.1200(b) of the OSHA regulations exempt from their requirements the following materials:

1. Any hazardous waste as that term is defined by the Solid Waste Disposal Act, as amended (42 U.S.C. Section 6901, *et seq.*), when subject to regulation under that act;

2. Tobacco or tobacco products;

3. Wood or wood products;

4. "Articles", which are defined as manufactured items which are:

    (a) formed to a specific shape or design during manufacture;

---

[5] *Id.*

[6] *Id.*

(b) which have end use functions dependent in whole or in part upon its shape or design during end use; and

(c) which do not release or otherwise result in exposure to a hazardous chemical under normal conditions of use.

5. Food, drugs, cosmetics, or alcoholic beverages in a retail establishment which are packaged for sale to consumers;

6. Foods, drugs, or cosmetics intended for personal consumption by employees while in the workplace;

7. Any consumer product or hazardous substance, as those terms are defined in the Consumer Product Safety Act[7] and the Federal Hazardous Substances Act,[8] respectively, where the employer can demonstrate that such material is used in the workplace in the same manner as in normal consumer use and that such use results in an exposure which is not greater than that experienced by consumers in terms of duration and frequency; and

8. Any drug, as that term is defined in the Federal Food, Drug and Cosmetic Act,[9] when it is in its final form for direct administration to the patient.[10]

In addition, Section 311(e) of EPCRA excludes the following substances:

1. Any food, food additive, color additive, drug, or cosmetic regulated by the Food and Drug Administration;

---

[7] 15 U.S.C. § 2051, et seq.

[8] 15 U.S.C. § 1261, et seq.

[9] 21 U.S.C. § 301, et seq.

[10] 29 C.F.R. § 1910.1200(b)(6).

2. Any substance present as a solid in any manufactured item to the extent exposure to the substance does not occur under normal conditions of use;

3. Any substance to the extent it is used for personal, family, or household purposes, or is present in the same form and concentration as a product packaged for distribution and use by the general public;

4. Any substance to the extent it is used in a research laboratory or a hospital or other medical facility under the direct supervision of a technically qualified individual;

5. Any substance to the extent it is used in routine agricultural operations or as a fertilizer held for sale by a retailer to the ultimate customer.[11]

Section 311 reporting is required for hazardous chemicals present at the facility, as of the date the report is due, in quantities equal to or greater than 10,000 pounds. For "extremely hazardous substances," *i.e.*, the substances listed under Section 302 and subject to emergency release notification under Section 304 of Subtitle A, which are *per se* a subset of the universe of hazardous substances, the reporting threshold is set at 500 pounds, 55 gallons, or the threshold planning quantity, whichever is less. EPA has extended the applicability of these thresholds through March 17, 1990, and has announced its intention to set final reporting thresholds probably in excess of the current zero quantity before September 1990.[12]

---

[11] § 311(a)(1), 42 U.S.C. § 11021(a)(1); 40 C.F.R. § 370.21.

[12] See 54 Fed. Reg. 41904-05 (October 12, 1989). Note also that EPA has proposed changing the dates upon which final
Continued

A facility owner or operator may submit an MSDS for each hazardous substance present at a facility in excess of the reporting threshold to the LEPC, SERC, and the fire department with jurisdiction over the facility. Alternatively, the owner or operator may submit a list of hazardous chemicals present in excess of the reporting threshold for which MSDSs are maintained.[13]/ If an owner or operator elects to provide a list of hazardous substances rather than the MSDSs themselves, the list must contain the following three types of information:

1. The hazardous substances must be grouped into the five specified hazard categories;[14]/

2. The chemical or common name of each hazardous chemical

Section 311 reports are due. For non-manufacturing employers other than the construction industry the Agency has proposed a final reporting date of October 17, 1990, and for the construction industry a date of October 17, 1990. See 54 Fed. Reg. 41907-08 (October 12, 1989).

[13]/ § 311(a)(2), 42 U.S.C. § 11021(a)(2); 40 C.F.R. § 370.21(b).

[14]/ EPA has compressed the 23 OSHA hazard categories into five. These categories are:

(1) "Immediate (acute) health hazard" (which includes the OSHA categories "highly toxic," "toxic," "irritant," "sensitizer," and "corrosive");

(2) "Delayed (chronic) health hazard" (which includes carcinogens);

(3) "Fire hazard" (which includes the OSHA categories "flammable," "combustible liquid," "pyrophoric," and "oxidizer");

(4) "Sudden release of pressure" (which includes the categories "explosive" and "compressed gas"); and

(5) "Reactive" (which includes the OSHA categories "unstable reactive," "organic peroxide," and "water reactive").

§ 311(a)(2)(A)(i), 42 U.S.C. § 11021(a)(2)(A)(i); 40 C.F.R. § 370.21(b)(1) and 40 C.F.R. § 370.2.

as provided on the MSDS;[15] and

3. An identification of any hazardous component of the substance as provided on the MSDS.[16]

Additionally, upon request of an LEPC, an owner or operator who submitted a list rather than an MSDS for hazardous chemicals at a facility must, within thirty days of receipt of such a request, submit a copy of the MSDS to the local committee.[17] Any person may request a copy of an MSDS from the LEPC, and if a local committee does not have the requested MSDS in its possession, it is obligated to request it from a facility owner or operator and provide it to the requester.[18]

If an owner or operator of a facility that has submitted an MSDS to satisfy the requirements of Section 311 comes into possession of significant new information concerning the hazardous chemical for which the MSDS was submitted, he must provide a revised MSDS to the LEPC, SERC, and local fire department within three months of the discovery of such information.[19] Similarly, within three months of the time a facility owner or operator is first required to prepare or have an MSDS available, or after a hazardous chemical requiring an MSDS first becomes present in an amount exceeding the threshold

---

[15] § 311(a)(2)(A)(ii), 42 U.S.C. § 11021(a)(2)(A)(ii); 40 C.F.R. § 370.21(b)(2).

[16] § 311(a)(2)(A)(iii), 42 U.S.C. § 11021(a)(2)(A)(iii); 40 C.F.R. § 370.21(b)(3).

[17] § 311(c)(1), 42 U.S.C. § 11021(c)(1); 40 C.F.R. § 370.21(d).

[18] § 311(c)(2), 42 U.S.C. § 11021(c)(2); 40 C.F.R. § 370.30(a).

[19] § 311(d)(2), 42 U.S.C. § 11021(d)(2); 40 C.F.R. § 370.21(c).

reporting quantity, the owner or operator must submit an MSDS or a listing for that chemical to the LEPC, SERC, and local fire department.[20]

Finally, because mixtures are included within OSHA's definition of chemical,[21] MSDSs have been created for tens of thousands of products. In recognition of this fact, two options have been provided for reporting on mixtures. An owner or operator may submit an MSDS or make a listing entry for the mixture itself.[22] If this alternative is chosen, for purposes of determining whether the mixture triggers the reporting threshold quantity the total quantity of the mixture must be reported.[23] Alternatively, an MSDS or listing entry may be submitted for each element or compound in the mixture which is itself a hazardous chemical.[24] To determine the reporting threshold quantity in this case, EPA has adopted a *de minimis* rule. Thus, for each component of a mixture that is a hazardous chemical present at more than one percent by weight (or 0.1

---

[20] § 311(d)(1)(B), 42 U.S.C. § 11021(d)(1)(B); 40 C.F.R. § 370.21(c)(2).

[21] "'Chemical' means any element, chemical compound, or mixtures of elements and/or compounds." 29 C.F.R. § 1910.1200(c).

[22] § 311(a)(3)(B), 42 U.S.C. § 11021(a)(3)(B); 40 C.F.R. § 370.28(a)(2).

[23] 40 C.F.R. § 370.28(b)(2).

[24] § 311(a)(3)(A), 42 U.S.C. § 11021(a)(3)(A); 40 C.F.R. § 370.28(a)(1). Note that under this option multiple reporting on the same hazardous chemical is not required. Thus, once an owner or operator has listed or submitted an MSDS for a particular hazardous chemical that is part of a mixture, it is not necessary to submit an additional MSDS or listing entry for that substance if it is part of another compound or mixture.

percent if the chemical is a carcinogen), that weight percentage is to be multiplied by the mass (in pounds) of the total mixture to determine the quantity of the hazardous chemical present in the mixture.[25] Finally, if a mixture is hazardous but contains no compounds or elements that are themselves hazardous, the owner or operator must submit an MSDS for the mixture.[26]

B. Section 312 -- Emergency and Hazardous Chemical Inventory Forms

Section 311 provides emergency planning and response authorities with information concerning the presence and characteristics of hazardous chemicals present at a facility. The "Emergency and Hazardous Chemical Inventory Forms" mandated by Section 312 supplement this information by requiring owners or operators to provide data on the quantities of such materials and their locations at reporting facilities. Facility owners or operators who submitted MSDSs or lists of hazardous chemicals for which they prepare or maintain MSDSs under Section 311 are required to submit inventory forms under Section 312.[27] Inventory forms are to be submitted to the same entities that received MSDSs or lists under Section 311, *i.e.*, the LEPC, SERC, and fire department with jurisdiction over the facility.[28]

Section 312 reporting is an annual obligation. Section 312

---

[25] 40 C.F.R. § 370.28(b).

[26] H.R. Conf. Rep. No. 962, 99th Cong., 2d Sess. 287 (1986).

[27] § 312(a)(1), 42 U.S.C. § 11022(a)(1); 40 C.F.R. § 370.20(a).

[28] §§ 312(a)(1)(A)-(C), 42 U.S.C. §§ 11022(a)(1)(A)-(C); 40 C.F.R. § 370.25(a).

reports are due annually on March 1 of each succeeding year.[29] The same quantity-based reporting thresholds that apply to Section 311 also apply to Section 312 inventory reporting.[30]

Section 312 creates a two-tiered reporting system under which Tier I information is to be submitted annually.[31] Tier II information may be submitted in lieu of Tier I data,[32] however, and upon request of the LEPC, SERC, or local fire department, an owner or operator must submit Tier II data within thirty days of receipt of such a request.[33]

The statute specifies the types of information that are to be provided in Tier I and Tier II reports, and EPA has promulgated forms for these submissions.[34] EPCRA requires Tier

---

[29] § 312(a)(2), 42 U.S.C. § 11022(a)(2); 40 C.F.R. § 370.20(b)(2).

[30] § 312(b), 42 U.S.C. § 11022(b); 40 C.F.R. § 370.20(b)(2).

[31] § 312(a)(2), 42 U.S.C. § 11022(a)(2); 40 C.F.R. § 370.25(a). This tiering system was developed in an effort:

> [t]o minimize the burden of this reporting and to provide the information in a manner which is of maximum usefulness to government emergency response offices and personnel, other government officials with a need for the information, and to the public....

H.R. Conf. Rep. No. 962, 99th Cong., 2d Sess. 288 (1986).

[32] § 312(a)(2), 42 U.S.C. § 11022(a)(2); 40 C.F.R. § 370.25(b).

[33] § 312(e), 42 U.S.C. § 11022(e); 40 C.F.R. § 370.25(c).

[34] § 312(d)(1), 42 U.S.C. § 11022(d)(1); 40 C.F.R. § 370.40 (Tier I), and § 312(d)(2), 42 U.S.C. § 11022(d)(2); 40 C.F.R. § 370.41 (Tier II). EPA promulgated Tier I and Tier II reporting forms in its October 15, 1987 rulemaking (52 Fed. Reg. 38344). The regulations provide, however, that an owner or operator may submit Tier I and Tier II data on state or local forms provided that such forms' contents are identical to the Federal forms'. 40 C.F.R. §§ 370.40 and 370.41.

I reports to contain three basic categories of information which are to be presented in the aggregate with hazardous chemicals grouped into health and physical hazard categories.35/ The information required includes:

1. An estimate, in ranges, of the maximum amount of hazardous chemicals in each of the five hazard categories present at the facility at any time during the preceding calendar year;

2. An estimate, in ranges, of the average daily amount of hazardous chemicals in each of the five hazard categories present at the facility during the preceding calendar year; and

3. The general location of hazardous chemicals in each of the five hazard categories at the facility.36/

Unlike Tier I, Tier II reports are chemical-specific, i.e., data are required for individual hazardous chemicals rather than for hazard categories.37/ As with the contents of Tier I reports, the data required for Tier II submissions are specified by statute. These data are:

1. The chemical or common name of the hazardous chemical as given on the MSDS;

2. An estimate, in ranges, of the greatest amount of the substance present at the facility during the preceding calendar year;

35/ These hazard categories are the same as those established for Section 311 reporting, i.e., fire hazard, sudden release of pressure, reactivity, immediate (acute) health hazard, and delayed (chronic) health hazard. See Note 68, supra.

36/ § 312(d)(1)(B), 42 U.S.C. § 11022(d)(1)(B); 40 C.F.R. § 370.40.

37/ § 312(d)(2), 42 U.S.C. § 11022(d)(2).

3. An estimate, in ranges, of the average daily amount of the substance present at the facility during the preceding calendar year;

4. A brief description of the manner in which the substance is stored;

5. The location of the substance at the facility; and

6. Whether the owner or operator wishes to have the storage location of the substance withheld from public disclosure.[38]

Note that item six above represents a deviation from EPCRA's blanket prohibition on confidentiality claims for information other than specific chemical identity. An owner or operator who wishes to have the storage location withheld from public disclosure need not submit a substantiation for that trade secrecy claim; however, if the owner or operator wishes to claim the specific chemical identity as a trade secret, he must submit the Federal Tier II form to EPA along with appropriate substantiation for that claim in accordance with the Agency's regulations implementing Sections 322 and 323.[39]

As noted, the same reporting thresholds that apply under Section 311 apply under Section 312. Likewise, the same provisions that apply to reporting on mixtures containing hazardous substances under Section 311 apply to inventory forms

---

[38] §§ 312(d)(2)(A)-(F), 42 U.S.C. §§ 11022(d)(2)(A)-(F); 40 C.F.R. § 370.41.

[39] See 52 Fed. Reg. 38355 (Oct. 15, 1987).

submitted under Section 312.[40] To the extent practicable, the manner of reporting on mixtures under Section 311 is to be followed by an owner or operator in Section 312 reports.[41]

Finally, EPCRA contains a number of explicit provisions concerning the public dissemination of Tier II data. Not only is an owner or operator obligated to respond to a request by the LEPC, SERC, or local fire department for Tier II data,[42] but these data must be submitted irrespective of the quantity of the substance present at the facility, i.e., the reporting thresholds do not apply.[43] In addition, any person may submit a written request to a SERC or LEPC for Tier II information, and if the SERC or LEPC has such data in its possession, it must provide them. If it does not, it must request Tier II information from the appropriate facility and supply it to the person making the request for any hazardous chemical that was present at the facility in an amount in excess of 10,000 pounds.[44] If the SERC

---

[40] § 312(a)(3), 42 U.S.C. § 11022(a)(3); 40 C.F.R. § 370.28.

[41] 40 C.F.R. § 370.28(a)(2). See also H.R. Conf. Rep. No. 962, 99th Cong., 2d Sess. 289 (1986). The Conference Report is particularly illuminating with regard to reporting on mixtures. It indicates, for example, that if a mixture is hazardous, although its constituents per se are not, an inventory form must be submitted. It also observes that if an owner or operator submits inventory data on each element or compound in the mixture, he may aggregate the amount of such element or compound present at the facility in its pure state with the amount in all mixtures at the facility as a single inventory entry for the amount of that compound or element present at the facility. Id.

[42] § 312(e)(1), 42 U.S.C. § 11022(e)(1); 40 C.F.R. § 370.25(c).

[43] 40 C.F.R. § 370.20(b)(3).

[44] §§ 312(e)(3)(A) and (B), 42 U.S.C. §§ 11022(e)(3)(A) and (B); 40 C.F.R. § 370.30(b).

or LEPC receives a request for Tier II information not in its possession which concerns a substance present at a facility in an amount less that 10,000 pounds, the regulatory authority is given discretion to decide whether or not to request such information from a facility.[45]

    C. **Section 313 -- Toxic Chemical Release Reporting**

The third and final reporting provision of Subtitle B is the Section 313 toxic chemical release report submitted annually to EPA and to a designated state official[46] beginning on July 1,

---

[45] § 312(e)(3)(C), 42 U.S.C. § 11022(e)(3)(C); 40 C.F.R. § 370.30(b)(3). Additionally, the person making such a request must include with the request a statement of the general need for such information.

The different treatment of data concerning hazardous substances present in quantities of more or less than 10,000 pounds appears to have evolved from a congressional balancing of the public's need to know and owner's or operator's concerns to protect trade secrets from competitors. This is suggested by the following admonitory passage in the Conference Report:

> Although the conference substitute establishes a procedure for the public to have access to this information, and business establishments are certainly a part of "the public," this provision is not intended to provide a means for competitors to find out confidential business information about each other. State emergency response commissions and local emergency planning committees should exercise their discretion in light of this consideration.

H.R. Conf. Rep. No. 962, 99th Cong., 2d Sess. 291 (1986).

[46] § 313(a), 42 U.S.C. § 11023(a); 40 C.F.R. § 372.30(d).

-135-

1988, and by each July 1 thereafter. Section 313 requires reporting on releases[47] into each environmental medium of specified "toxic chemicals" that can reasonably be anticipated to cause adverse human health effects or significant adverse effects on the environment.[48] EPA promulgated the final Section 313 reporting regulations and reporting form on February 16, 1988.[49]

The statute establishes threshold criteria that are used to determine a facility's obligation to report. The first of these is that the facility must have ten or more full-time employees.[50] Second, the facility must be grouped in SIC Codes 20 through 39.[51] Third, the facility must manufacture,[52] process,[53] or otherwise use a toxic chemical in excess of the

---

[47]   See Note 42, supra.

[48]   H.R. Conf. Rep. No. 962, 99th Cong., 2d Sess. 292 (1986).

[49]   53 Fed. Reg. 4500.

[50]   § 313(b)(1)(A), 42 U.S.C. § 11023(b)(1)(A); 40 C.F.R. § 372.22(a). The term "full-time employee" is defined to mean "2,000 hours per year of full-time equivalent employment." 40 C.F.R. § 372.3 To determine the number of full-time employees a facility has, the total number of hours worked by all employees, including contract employees, is totalled and that number is divided by 2,000. Id.

[51]   § 313(b)(1)(A), 42 U.S.C. § 11023(b)(1)(A); 40 C.F.R. § 372.22(b). EPA interprets this requirement to relate to the primary SIC Code of the facility. If the facility comprises more than one establishment, the applicability of Section 313 is based on a relative comparison of the value of products shipped and/or produced at the establishments within Codes 20 through 39 as opposed to those grouped in other SIC Codes. Id. See also 53 Fed. Reg. 4501-4504.

[52]   § 313(b)(1)(C)(i), 42 U.S.C. § 11023(b)(1)(C)(i); 40 C.F.R. § 372.3.

[53]   § 313(b)(1)(C)(ii), 42 U.S.C. § 11023(b)(1)(C)(ii).

established reporting thresholds. The statute defines the term "manufacture" to mean: "produce, prepare, import, or compound a toxic chemical." In its regulations, however, EPA has expanded upon this definition and has construed the term "manufacture" to mean:

> to produce, prepare, import, or compound a toxic chemical. Manufacture also applies to a toxic chemical that is produced coincidentally during the manufacture, processing, use or disposal of another chemical or mixture of chemicals, including a toxic chemical that is separated from that other chemical or mixture of chemicals as a by-product, and a toxic chemical that remains in that other chemical or other mixture of chemicals as an impurity.[54]

EPA has taken this step because of its belief that the statutory definition of the term manufacture includes the coincidental production of toxic chemicals. For reporting purposes, however, EPA distinguishes between toxic chemicals that remain as impurities in another chemical from toxic chemicals that are by-products which are either sent for disposal or processed, distributed, or used in their own right. Thus, if a toxic chemical is produced coincidentally as a by-product in excess of the reporting threshold, reporting will be required. However, if the toxic chemical is present in another chemical or mixture of chemicals as an impurity, EPA applies a de minimis rule for determining threshold reporting quantities. Thus if a

---

[54] 40 C.F.R. § 372.3.

toxic chemical is present as an impurity in a concentration of one percent (0.1% if the toxic chemical is a carcinogen) or less, the quantity of that chemical need not be considered for purposes of determining whether a reporting threshold has been met and reporting is required.[55]

The statute defines the term "process" to mean:

> the preparation of a toxic chemical after its manufacture for distribution in commerce -
>
> (I) in the same form or physical state as, or in a different form or physical state from, that in which it was received by the person so preparing such chemical, or
>
> (II) as part of an article containing the toxic chemical.

EPA has expanded part (II) of this definition to include "the processing of a toxic chemical contained in a mixture or trade name product."[56]

In essence, the term process means to prepare a toxic chemical after its manufacture for distribution in commerce. Processing includes a variety of activities, such as blending, mixing, changing the physical state of the chemical, and incorporating the chemical into an article. A material is processed if, after manufacture, it is made part of a material or product distributed in commerce. Note that a toxic chemical can be processed if it is part of a mixture or trade name product,

---

[55] 40 C.F.R. § 372.28; see also 53 Fed. Reg. 4504-4505.
[56] 40 C.F.R. § 372.3.

including such a chemical that is present as an impurity in a product. If a material containing an impurity is processed, that impurity is being processed, and if it is a toxic chemical and exceeds the reporting threshold, it must be reported.[57]

A toxic chemical that is "otherwise used" is one in which the chemical does not deliberately become a part of the product that is distributed in commerce, for example, toxic chemicals used as catalysts, solvents, or reaction terminators.[58] EPA also gives as examples, lubricants, refrigerants, metal working fluids, or chemicals used for other purposes, such as cleaners, degreasers, or fuels.[59]

Reporting thresholds are specified by the statute and require a report to be filed:

(1) for a toxic chemical *used* at the facility in a quantity of at least 10,000 pounds during the preceding calendar year;

(2) for a toxic chemical *manufactured* or *processed* at a

---

[57] 40 C.F.R. § 372.3. See 53 Fed. Reg. 4505-4506.

[58] EPCRA does not define the term "otherwise used". EPA, however, has defined the term to mean:

> any use of a toxic chemical that is not covered by the terms "manufacture" or "process" and includes use of a toxic chemical contained in a mixture or trade name product. Relabeling or redistributing a container of a toxic chemical where no repackaging of the toxic chemical occurs does not constitute use or processing of the toxic chemical.

40 C.F.R. § 372.3.

[59] See 53 Fed. Reg. 4505-4506.

facility in a quantity of at least 25,000 pounds during the preceding calendar year for the reports due July 1, 1990 and for each year thereafter.[60] Note that if more than one threshold applies to activities involving a toxic chemical at a facility, i.e., if it is both manufactured and otherwise used, the owner or operator must report on all activities at the facility involving the chemical if any one of the reporting thresholds is exceeded.[61] Also, if a facility reuses or recycles a toxic chemical on-site, to determine if the reporting threshold for that chemical has been exceeded, the owner or operator must determine how much of the chemical was added to the re-use/recycle system during the preceding year.[62]

Another point of interaction between the definitions and threshold reporting quantities should be noted. This is that while the terms "process" and "otherwise used" do not relate to the amount of toxic chemical brought on-site, the term "manufacture" does. Thus, if an owner or operator brings a chemical on-site via importation in excess of the reporting threshold for the manufacture of a toxic chemical, all emissions of that chemical must be reported.[63]

Subject to a one percent (0.1 percent for carcinogens) de

---

[60] § 313(f)(1), 42 U.S.C. § 11023(f)(1); 40 C.F.R. § 372.25.
[61] 40 C.F.R. § 372.25(c).
[62] 40 C.F.R. § 372.25(e). See also 53 Fed. Reg. 4508.
[63] 40 C.F.R. § 372.25(e).

minimis concentration limit, release reports are also required for toxic chemicals that are components of trade name products that an owner or operator imports, processes, or otherwise uses, if the quantity of such chemicals exceeds the applicable reporting thresholds. The key to this requirement is the stipulation in the statute that release reports are required for toxic chemicals exceeding the threshold reporting limits for "each listed toxic chemical known to be present at the facility."[64]/ An owner or operator has knowledge of the presence of a toxic chemical in a mixture or trade name product if he knows or has been told the identity or Chemical Abstracts Service ("CAS") Registry Number of a substance, and that identity or CAS number corresponds to a toxic chemical listed in 40 C.F.R. § 372.65. An owner or operator also has knowledge if he has been told by the supplier of the mixture or trade name product that the substance contains a toxic chemical subject to Section 313 of EPCRA.[65]/

The regulations provide a number of reporting alternatives for mixtures or trade name products based upon the amount of information available to the owner or operator. For example, if the owner or operator knows both the specific chemical identity of the toxic chemical and its specific concentration in the mixture or the trade name product, he is required to determine

---

[64]/ § 313(g)(1)(C), 42 U.S.C. § 11023(g)(1)(C).

[65]/ 40 C.F.R. § 372.30(a)(2). See also 53 Fed. Reg. 4508-4511.

the weight of the chemical in the mixture or trade name product and to combine that with the weights of that chemical manufactured, processed, or otherwise used at the facility for purposes of determining the reporting threshold.[66/]

If the owner or operator knows the specific chemical identity and does not know its exact concentration but has been told the upper bound of its concentration in a mixture or trade name product, he is required to assume that the substance is present in the mixture or trade name product at the upper bound concentration and is to use that value for determining whether a threshold reporting quantity has been met.[67/]

If the owner or operator knows the specific chemical identity and knows neither the specific concentration nor the upper bound concentration and has not otherwise developed information on the composition of the chemical, then he is not required to factor that chemical in the mixture or trade name product into his threshold reporting quantity calculations.[68/]

If the owner or operator has been told that a mixture or trade name product contains a toxic chemical, does not know the specific chemical identity, and knows the specific concentration of the substance, he is required to determine the weight of the chemical as part of the mixture or trade name product. If a reporting threshold is met, the owner or operator is required to

---

[66/] 40 C.F.R. § 372.30(b)(3)(i).

[67/] 40 C.F.R. § 372.30(b)(3)(ii).

[68/] 40 C.F.R. § 372.30(b)(3)(iii).

report using the generic chemical name of the toxic chemical or its trade name, if the generic chemical name is not known.[69]/

If the owner or operator has been told that a mixture or trade name product contains a toxic chemical, does not know the specific chemical identity or the specific concentration, but has been told an upper bound concentration, he is required to assume that the toxic chemical is present at the upper bound concentration and, if the reporting threshold is crossed, to report giving the generic name or trade name.[70]/

If the owner or operator has been told that a mixture or trade name product contains a toxic chemical, does not know the specific chemical identity, does not know specific concentration, and does not know the upper bound concentration, the owner or operator is not required to report on that chemical.[71]/

To facilitate such reporting, EPA has imposed supplier notification requirements.[72]/ These requirements apply to facility owners or operators in SIC Codes 20 through 39 who manufacture or process products containing toxic chemicals which they sell or distribute in mixtures or trade name products to facilities subject to reporting under Section 313.[73]/ This

---

[69]/ 40 C.F.R. § 372.30(b)(3)(iv).

[70]/ 40 C.F.R. § 372.30(b)(3)(v).

[71]/ 40 C.F.R. § 372.30(b)(3)(vi).

[72]/ 40 C.F.R. § 372.45.

[73]/ 40 C.F.R. § 372.45(a).

notice, which is to be attached to an MSDS, if one is required, must contain certain specified elements of information, generally including:

(1) A statement that the mixture or trade name product contains a toxic chemical subject to Section 313 reporting;

(2) The name and CAS number for each such toxic chemical;

(3) The percent, by weight, of each toxic chemical in the mixture or trade name product.[74]

Notification is not required if the concentration of a toxic chemical in a mixture or trade name product does not exceed the applicable *de minimis* concentration, or if the mixture or trade name product is an article,[75] food, drug, cosmetic, alcoholic beverage, tobacco, or tobacco product packaged for distribution for general public, or a consumer product as defined by the Consumer Product Safety Act.[76]

Supplier notifications may also vary in content. For example, if the supplier claims the specific chemical identity of

---

[74] 40 C.F.R. § 372.45(b).

[75] "Article" is defined to mean:

> a manufactured item (1) which is formed to a specific shape or design during manufacture; (2) which has end use functions dependent in whole or in part upon its shape or design during end use; and (3) which does not release a toxic chemical under normal conditions of processing or use of that item at the facility or establishments.

40 C.F.R. § 372.3.

[76] 40 C.F.R. § 372.45(d).

a toxic chemical in a mixture or trade name product to be a trade secret (as defined by the OSHA hazard communication rules (29 C.F.R. § 1910.1200)), the notice must contain a generic chemical name.[77] Similarly, if the supplier considers the specific percent-by-weight composition to be a trade secret under state law or under Section 757, comment b, of the Restatement of Torts, the notice must provide an upper bound concentration value.[78] These requirements do not apply to an owner or operator who does not know if his facility is selling or otherwise distributing a toxic chemical in a mixture or trade name product to another person. If that owner or operator receives a supplier notification from his upstream supplier, however, he is deemed to have such notice.[79] These requirements take effect with the first product shipment in 1989.[80]

EPA has published a toxic chemical release inventory reporting form, the so-called EPA Form R.[81] This form is to be used for reporting on the following six types of information demanded by EPCRA:[82]

(1) The name, location, and principal business activities

---

[77] 40 C.F.R. § 372.45(e).

[78] 40 C.F.R. § 372.45(f).

[79] 40 C.F.R. § 372.45(g).

[80] 40 C.F.R. § 372.45(c).

[81] 53 Fed. Reg. 4500 (Feb. 16, 1988).

[82] § 313(g)(1), 42 U.S.C. § 11023(g)(1); 40 C.F.R. § 372.85. See also 53 Fed. Reg. 4511-4518.

of the facility. EPA has expanded this statutory requirement to include the latitude and longitude of the reporting facility, the facility's Dunn & Bradstreet number and, if applicable, the facility's parent company's Dunn & Bradstreet number, the names of technical and public contact persons, RCRA identification number, NPDES permit number, and underground injection well code. The Agency also wants reported the names of each stream or surface water body receiving toxic chemical discharges from the facility and identification of any off-site treatment, storage, or disposal facilities and POTW to which wastes containing toxic chemicals from the facility were sent.

(2) The submission must be certified as accurate by "a senior official with management responsibility for the person or persons completing the report, regarding the accuracy and completeness of the report.[83] EPA has construed this requirement broadly, and has defined the term "senior management official" to mean:

> an official with management responsibility for the person or persons completing a report, or the manager of environmental programs for the facility or establishments, or for the corporation owning or operating the facility or establishments responsible for certifying similar reports under other environmental regulatory requirements.[84]

(3) For each listed toxic chemical known to be present at the facility in excess of the applicable threshold, the report is

---

[83] § 313(g)(1)(B), 42 U.S.C. § 11023(g)(1)(B).
[84] 40 C.F.R. § 372.3.

to provide the chemical name and, if applicable, the CAS number of the substance and to contain the following items of information:[85]

(a) Whether the toxic chemical is manufactured, processed, or otherwise used at the facility and the general category or categories of such use;

(b) An estimate, in ranges, of the maximum amount of a toxic chemical present at the facility at any time during the preceding calendar year;

(c) Identification of the waste treatment or disposal method used by the facility for each waste stream containing any listed toxic chemical and "an estimate of the treatment efficiency typically achieved by such methods for that waste stream."[86] Waste streams treated in the same manner are aggregated. For example, all the wastes going to a secondary waste water treatment system on site would be combined in the report rather than each individual waste stream being reported separately. If some waste streams containing the toxic chemical are treated separately, however, then the owner or operator must

---

[85] Note that the specific chemical identity can be claimed to be a trade secret in accordance with Section 322 of EPCRA. If an owner or operator wishes to make such a claim, he must supply a generic chemical name in the report, and submit a substantiation of the claim to EPA along with the report. He must also submit a non-confidential copy of the report to EPA and to the state official designated to receive the report. EPA has published a form for the submission of substantiations accompanying trade secrecy claims. See 40 C.F.R. Part 350, 53 Fed. Reg. 28772 (July 29, 1988).

[86] § 313(g)(1)(C)(iii), 42 U.S.C. § 11023(g)(1)(C)(iii).

report on them individually. The term "treatment efficiency" refers to the mass percent by which the subject toxic chemical is removed from the waste stream, not to overall system efficiency. Removal in this context includes destruction, chemical conversion, physical removal, or some combination thereof. The report also calls for submission of a ranged estimate of the concentration of the toxic chemical in the influent to the waste treatment system;[87]/ and

(d) The quantity of each toxic chemical entering each environmental medium annually. This requirement includes the off-site shipment of waste, including discharges to POTWs. In the reports due for calendar years 1987-1989, releases via off-site transfers of quantities of less than 1,000 pounds may be submitted in ranges. EPA has construed this requirement to apply to total annual releases to environmental media, including both routine and accidental releases.[88]/

The statute provides that reports under Section 313 may be based on "reasonably available data", including monitoring and emission measurements collected pursuant to other environmental statutes, and that if no such data exist, reasonable estimates may be used.[89]/ Form R requires that these estimates be rounded

---

[87]/ See 53 Fed. Reg. 4516-4519.

[88]/ See 53 Fed. Reg. 4513.

[89]/ In order to provide the information required under this section, the owner or operator of a facility may use readily available data (including monitoring data) collected

Continued

to no more than two significant digits and that the basis from which they are derived (e.g., monitoring data, mass balance data, emission factors, or best engineering judgment) be indicated. Also, EPA has required that data on releases to the various environmental media be disaggregated: for example, air emissions must be identified as fugitive or point source, releases to water must be separated by receiving stream (as noted previously, waste water sent to a POTW is identified as an off-site transfer).[90]/

Finally, there are several exemptions from the Section 313 reporting requirement. Two of these, the _de minimis_ concentration and articles exemptions, have been discussed previously. Another exemption applies to certain uses of toxic chemicals. Uses which are exempt include toxic chemicals used as structural components of a facility, products used for routine janitorial or facility maintenance (e.g., fertilizers and pesticides), toxic chemicals used for personal purposes by employees or other persons at the facility as foods, drugs, cosmetics, or other personal items (including products within the

---

          pursuant to other provisions of law, or where such data are not readily available, reasonable estimates of the amounts involved. Nothing in this section requires the monitoring or measurement of the quantities, concentration, or frequency of any toxic chemical released into the environment beyond that monitoring and measurement required under other provisions of law or regulation.

§ 313(g)(2), 42 U.S.C. § 11023(g)(2).

90/ See 53 Fed. Reg. 4513-4516.

facilities in a cafeteria, store, or infirmary), products containing toxic chemicals used for maintaining motor vehicles, and toxic chemicals contained in process water and noncontact cooling water or in air used as compressed air or for combustion.[91]

Similarly, toxic chemicals otherwise subject to Section 313 reporting that are manufactured, processed, or used in a laboratory at a facility under the supervision of a technically qualified individual need not be considered in determining whether a threshold quantity of the toxic chemical is present at the facility. This exemption, however, does not apply to specialty chemical production, pilot plant operations, or activities conducted outside the laboratory.[92]

Note that a person who owns a facility otherwise subject to Section 313 reporting need not report if his only interest in the facility is ownership of the real estate upon which the facility is operated. This applies to owners of such facilities as industrial parks. Also, if two or more persons, without any common corporate or business interests, operate separate establishments within a single facility, each such person must report separately to the extent that person is covered by the requirements of Section 313.[93]

---

[91] 40 C.F.R. §372.38(c)(1)-(5).

[92] 40 C.F.R. §§ 372.38(d)(1)-(3).

[93] 40 C.F.R. §§ 372.38(e) and (f).

EPCRA authorizes EPA, <u>sua sponte</u>, or upon the petition of a person or state governor, to initiate rulemaking to add or delete substances from the list of toxic chemicals.[94] A chemical may be added to the list if sufficient evidence supports a finding that: (1) the toxic chemical causes or may reasonably be anticipated to cause significant adverse acute human health effects; (2) it causes or can reasonably be anticipated to cause various chronic human health effects; or (3) because of its toxicity, environmental persistence, or tendency to bioaccumulate,[95] the toxic chemical causes or can reasonably be anticipated to cause significant adverse environmental effects.[96]

---

[94] §§ 313(d) and (e), 42 U.S.C. §§ 11023(d) and (e).

[95] The term "bioaccumulate" is not used in a specific, technical sense in the statute. It "is not intended to distinguish between this term [bioaccumulate] and other technical terms such as 'bioconcentrate' and 'biomagnify' that sometimes are used interchangeably." H.R. Conf. Rep. No. 962, 99th Cong., 2d Sess. 295 (1986).

[96] § 313(d)(2), 42 U.S.C. § 11023(d)(2).

Acute health effects are those reasonably likely to exist beyond a facility boundary as a result of continuous or frequently recurring releases. <u>Id.</u> In determining whether to list a particular chemical because of its propensity to cause acute human health effects, EPA is directed to consider a variety of factors, <u>e.g.</u>, volume and pattern of use or release and individuals who are sensitive to that chemical. H.R. Conf. Rep. No. 962, 99th Cong., 2d Sess. 294 (1986). The term "continuing or frequently recurring releases" is used to distinguish releases that are a normal part of a facility's operations from "episodic and accidental" releases that are subject to emergency reporting under Section 304. <u>Id.</u>

The Agency is also authorized to modify the reporting thresholds.[97/] It may also modify the frequency of reporting,

---

The statute identifies the following chronic health effects: (1) Cancer or teratogenic, or (2) serious or irreversible -- (a) reproductive dysfunctions, (b) neurological disorders, (c) effects of heritable genetic mutations, or (d) other chronic health defects. § 313(d)(2)(B), 42 U.S.C. § 11023(d)(2)(B). Note that although the statute speaks of such health effects in humans, EPA is not restricted in deciding whether to list a substance because of its potential to cause such effects to chemicals for which human data exists. H.R. Conf. Rep. No. 962, 99th Cong., 2d Sess. 294 (1986).

Chemicals listed because of adverse environmental effects may not exceed twenty-five percent of the total number of toxic chemicals listed. § 313(d)(2)(C)(iii), 42 U.S.C. § 11023(d)(2)(C)(iii).

The Conference Report lists a variety of factors EPA should consider in determining whether to list a substance under this subsection:

> (1) Gradual or sudden changes in the composition of animal life or plant life, including fungi or microbial organisms in any area;
>
> (2) Abnormal number of deaths of organisms (e.g., fish kills);
>
> (3) Reduction of the reproductive success or the vigor of a species;
>
> (4) Reduction in agricultural productivity, whether crops or livestock;
>
> (5) Alteration in the behavior or distribution of a species;
>
> (6) Long lasting or irreversible contamination of components of the physical environment, especially in the case of groundwater and surface water and soil resources that have limited self-cleaning capability.

H.R. Conf. Rep. No. 962, 99th Cong., 2d Sess. 295
Continued

although it may not require that such reports be submitted more frequently than once per year.[98]/ Modifications to the frequency of reporting must be made by rulemaking and be supported by substantial evidence.[99]/ EPA must find that a modification is consistent with the function of Section 313, i.e., "to inform persons about releases of toxic chemicals to the environment; to assist governmental agencies, researchers, and other persons in the conduct of research and data gathering; to aid in the development of appropriate regulations, guidelines, and standards; and for other similar purposes."[100]/

EPA is also authorized to add or delete SIC Codes from the Section 313 reporting obligation [101]/ and may, sua sponte, or at

---

(1986).

[97]/ So long as the resulting reports cover a substantial majority of the aggregate releases of a chemical at a facility. § 313(f)(2), 42 U.S.C. § 11023(f)(2).

[98]/ § 313(i)(1), 42 U.S.C. § 313(i)(1).

[99]/ §§ 313(i)(2)(B) and (i)(6), 42 U.S.C. §§ 11023(i)(2)(B) and (i)(6).

[100]/ § 313(h), 42 U.S.C. § 11023(h); H.R. Conf. Rep. No. 962, 99th Cong., 2d Sess. 299-300 (1986).

[101]/ § 313(b)(1)(B), 42 U.S.C. § 11023(b)(1)(B). EPA's authority in this regard is limited to adding or deleting codes to the degree needed to carry out the purposes of Section 313. That is, EPA may add SIC Codes:

> for facilities which, like the facilities within the manufacturing sector SIC Codes 20 through 39, manufacture, process, or use toxic chemicals in a manner such that reporting by these facilities is relevant to the purposes of this section. Similarly, the authority to delete SIC Codes from within SIC Codes 20 through 39 is limited to

Continued

the request of a governor, apply Section 313 reporting requirements to any particular facility that manufactures a toxic chemical substance.[102]/

The statute requires EPA to develop and maintain in a computerized data base a national toxic chemical inventory based upon data submitted in Section 313 reports.[103]/ These data are to be made available, at cost, to any person via computer telecommunications.[104]/

---

delisting SIC Codes for facilities which, while within the manufacturing sector SIC Codes, manufacture, process or use toxic chemicals in a manner more similar to facilities outside the manufacturing sector.

H.R. Conf. Rep. No. 962, 99th Cong., 2d Sess. 292 (1986).

[102]/ § 313(b)(2), 42 U.S.C. § 11023(b)(2). EPA may base this action upon consideration of such features as the toxicity of the toxic chemical, proximity of the facility to other facilities which release that substance or to population centers, and the history of releases of such substances from the facility. Id.

[103]/ § 313(j), 42 U.S.C. § 11023(j).

[104]/ Id.

THE TOXIC SUBSTANCES CONTROL ACT:
REPORTING AND RECORDKEEPING REQUIREMENTS

Charles A. O'Connor, III
Thomas B. Johnston

McKenna, Conner & Cuneo

I. INTRODUCTION

The Toxic Substances Control Act (TSCA)[1] contains extensive recordkeeping and reporting requirements to ensure that the EPA Administrator continually has access to new information developed regarding adverse health or environmental effects associated with chemical substances. The most significant reporting requirements are in § 8 of TSCA. Under § 8(a) EPA can require companies to collect, maintain and submit information on chemical production and processing; under § 8(c) EPA can require manufacturers and processors to maintain records of significant adverse reactions alleged to have been caused by their chemicals; under § 8(d) EPA can require companies to search their files and provide lists and copies of health and safety studies they have on their chemicals; and under § 8(e) companies must immediately report information concerning substantial risk.

A variety of other recordkeeping and reporting requirements are scattered throughout TSCA, such as (1) the Inventory update rule under § 8(b); (2) premanufacture notification under § 5; (3) R&D and test marketing under § 5(h); (4) consent orders under § 5(e) and significant new use rules under § 5(a); (5) biotechnology reporting; (6) reporting for PCBs and other chemicals regulated under § 6; (7) import and export notifications under §§ 13 and 12(b) respectively; and (8) reporting of asbestos removal in schools. The reporting requirements under § 8 and other sections of TSCA are discussed in more detail below.

II. TSCA SECTION 8: REPORTING AND RETENTION OF INFORMATION

TSCA § 8 establishes reporting and recordkeeping requirements to provide EPA with information on which to base regulatory and enforcement actions, and to track patterns of adverse reactions to chemicals. This section grants the Agency broad information-gathering authority. It was promulgated in response to a congressional perception that one of the most significant gaps in pre-TSCA environmental laws was the lack of authority for collecting data to determine the "totality of human and environmental exposure to chemicals." H.R. Rep. No. 1341, 94th Cong., 2d Sess. 6 (1976). Section 8 was designed to provide "an ongoing mechanism that would ensure that the EPA Administrator would continually have access to new information developed regarding adverse health or environmental effects associated with chemical substances." 122 Cong. Rec. S16,807 (daily ed. Sept. 28, 1976) (statement of Sen. Pearson).

Section 8(a) provides EPA with authority to promulgate rules requiring manufacturers to maintain records and make reports concerning the substances they produce, categories of uses, byproducts, environmental and health effects, and numbers of

---

[1] McKenna, Conner & Cuneo has produced a comprehensive treatise on the requirements of TSCA -- the TSCA Handbook (2d ed. 1989). Much of the material here is taken from the portions of the TSCA Handbook that address reporting and recordkeeping requirements; the TSCA Handbook provides comprehensive guidance on the full range of TSCA issues.

workers exposed. Section 8(b) requires EPA to publish and keep current an inventory of chemical substances. Section 8(c) requires manufacturers and certain processors to maintain records of adverse reactions to health or the environment. Section 8(d) requires manufacturers and processors to submit to EPA lists of health and safety studies and to provide copies of such studies upon request. Finally, § 8(e) requires manufacturers, processors, and distributors to inform EPA when they receive information that a chemical substance presents a substantial risk of injury to health or the environment.

EPA utilizes the information obtained under § 8 to provide information users such as other EPA programs, industry, and citizens with complete and reliable information on chemicals, to evaluate existing data to determine its adequacy for risk assessment purposes, to identify data gaps, and to monitor ongoing activities with respect to specific chemicals.

In order to accomplish these objectives, the Agency uses a number of approaches. First, EPA applies its § 8 authority as needed for specific chemicals and stops the reporting requirements when the information need ceases. Second, EPA's Office of Toxic Substances (OTS) coordinates its information-gathering activities with those of other programs and agencies. Finally, by using "model" rules and standardized forms, the Agency helps to reduce the cost of compliance to the regulated community, while minimizing possible reporting errors.

EPA has not exercised to any significant degree the sweeping information-gathering authority it possesses under TSCA § 8. Beginning in 1986, the Agency began to utilize § 8 with increasing frequency to support not only regulatory activities by OTS, but the regulatory activities of other EPA offices and other federal agencies as well. Also in 1986, EPA aggressively began to enforce all of the provisions of TSCA § 8, not just § 8(e). Enforcement actions have included the issuance of civil complaints against companies that allegedly violated the §§ 8(c) and 8(d) requirements.

In addition to its obvious relevance to EPA enforcement action, a company's compliance record also could have a bearing in the area of "toxic tort" liability. A growing number of suits have been filed by plaintiffs seeking redress from risk or injury caused by "toxic" substances. Failure to comply with TSCA's reporting and recordkeeping requirements could be cited by a plaintiff as evidence of negligence.

TSCA § 8(a): REPORTS

Under TSCA § 8(a), EPA may require companies to maintain records and submit reports on their chemical manufacturing, importing and processing activities. EPA uses information gathered under § 8(a) to determine what information exists on a chemical substance or mixture, and to set priorities for testing rules under TSCA § 4 and/or regulatory action under TSCA § 6. The Agency has used its § 8(a) authority to impose recordkeeping and reporting requirements on specific chemicals, but its most significant use of § 8(a) has been the issuance of "model" rules that apply to multiple chemicals. The first of these model rules, the Preliminary Assessment Information Rule (PAIR), was issued in June of 1982 and continues in effect. 47 Fed. Reg. 26,992 (1982) (codified at 40 C.F.R. pt. 712, subpt. B). The second, the Comprehensive Assessment Information Rule (CAIR), was issued on December 22, 1988 and, as explained below, will eventually replace the PAIR. 53 Fed. Reg. 51,698 (1988) (to be codified at 40 C.F.R. §§ 704.1-.225).

TSCA § 8(a) authorizes the Administrator to require those who currently manufacture or process chemical substances, or who propose to do so in the future, to maintain records and submit reports such as the Administrator "may reasonably require." TSCA § 8(a) authorizes the Administrator to require those who currently manufacture or process, or propose to manufacture or process, mixtures or small quantities of research and development (R&D) chemicals to maintain records and submit reports to the Agency only to the extent "necessary for the effective enforcement" of the Act. This same standard also applies to records and reports on changes in the proportions of the components of a mixture. The difference in these standards reflects Congress's belief that the Agency's need for information concerning mixtures and R&D chemicals is not as great as the need for information on basic chemical substances already in commerce. H.R. Rep. No. 1341, 94th Cong., 2d Sess. 41 (1976).

Small manufacturers and processors generally are exempt from § 8(a).[2/] The Administrator, however, can require reports from small manufacturers and processors of chemicals that are subject to test rules under § 4, appear on the "risk list" under § 5(b)(4), or are subject to limitations under §§ 5(e) or 6.

Section 8(a)(2) lists the kinds of information for which the Agency may require recordkeeping and reporting. Such information can include chemical identity, categories of use, amounts manufactured or processed, byproducts, data on environmental and health effects, and exposure information. Manufacturers and processors must provide this information to the extent they know or can reasonably ascertain it.

The Preliminary Assessment Information Rule (PAIR)[3/]

When EPA proposed the PAIR in 1980, the Agency viewed it as the first of a set of three progressively more detailed § 8(a) reporting rules that EPA would use to obtain information appropriate to the needs of initial, middle, and later stages of a comprehensive assessment of chemical substances. 45 Fed. Reg. 13,646 (1980). The two-page PAIR reporting form (EPA Form No. 7710-35) was intended to gather very preliminary exposure data to help the Agency set testing priorities on its master list of chemicals identified as potentially needing further testing and assessment due to their actual or potential toxicity.

As discussed below, the Agency has abandoned the concept of a set of progressively more detailed reporting rules in favor of one comprehensive reporting form (CAIR). At present, however, the PAIR continues in effect and any additional chemicals designated or recommended for testing by the Interagency Testing Committee (ITC) become subject to it automatically. The PAIR will continue to be used until such time as the Agency adds this automatic listing mechanism to the CAIR.

Who Must Report

Manufacturers and importers must report on each listed chemical substance manufactured or imported during the reporting period for that substance, as given in the rule. Processors are not subject to PAIR. Manufacturers of chemical substances extracted from natural sources, such as oil and ore, need only report the manufacturing steps and uses of the extracted chemical. They do not have to describe the extraction of the natural source material or other crude precursors derived from the natural source. 47 Fed. Reg. 26,993 (1982). For example, extraction of shale oil from a mine need not be reported, but the steps involved in retorting the ore and refining the shale oil into usable products must be reported.

[2/] TSCA does not define "small manufacturers" or "small processors." The PAIR and the CAIR define the terms differently, as discussed below.

[3/] The PAIR is implemented by the Chemical Screening Branch, Existing Chemicals Assessment Division (Tel. 202-382-3436).

EPA also exempts the following from reporting under the PAIR: "[1] Persons who manufacture or import a listed substance solely for R&D, including commercial R&D; [2] Persons who, during the reporting period, manufactured or imported fewer than 500 kilograms of a chemical at a single plant site; [3] Small manufacturers or importers, which the PAIR defines as companies whose total annual sales from all sites owned by the foreign or domestic parent company were below $30 million for the reporting period _and_ who produced less than 45,400 kg of the listed substance at the plant site;4/ [and 4] Persons who manufactured the chemical as a non-isolated intermediate, an impurity, or, under certain conditions, as a byproduct." 40 C.F.R. § 712.25.

Filling Out the Form

Manufacturers and importers subject to the PAIR must submit a two-page "Manufacturer's Report -- Preliminary Assessment Information" form for each plant site manufacturing a chemical listed in 40 C.F.R. § 712.30. Completed reports can be submitted by either the individual plants or the company headquarters.

The information must be provided to the extent it is "known to or reasonably ascertainable by" the respondent. EPA has interpreted this phrase to require "persons to report data that are readily obtainable by management and supervisory employees responsible for manufacturing, processing, distribution, technical services, marketing, and other related activities." 47 Fed. Reg. 26,994 (1982).

For each category of information, EPA has specified the accuracy level or range with which the information must be reported (e.g., ± 50 percent, "best estimates from readily obtainable data," etc.).

Certification By Respondent

The respondent must certify to the technical accuracy and confidentiality of the data reported. The physical location of the plant site and its mailing address must be identified along with any Dun and Bradstreet number. A technical contact person familiar with the information submitted also must be identified. EPA will send an acknowledgment of receipt of the form to this technical contact.

Chemical Identity

If the chemical has a Chemical Abstract Services (CAS) number and is listed in 40 C.F.R. § 712.30 by its specific chemical name, then the chemical substance must be identified on the form by its CAS number and the first fifteen characters of its name. If the chemical identity is considered confidential it is to be reported by the category name listed in 40 C.F.R. § 712.30 and the number of the Inventory Reporting Form C on which the chemical was originally reported for the TSCA Inventory.

Preliminary Assessment Information — Plant Site Activities

Information on plant site activities must be based on "best estimates from readily obtainable data." (Here, no specific accuracy range is required.) EPA has adopted this accuracy standard because the Agency was "persuaded that manufacturers must routinely know their own production efficiencies and quantities in order to control their costs and

---

4/ This exemption does not apply to manufacturers or importers of listed chemical substances that are designated as being of special concern. Such substances are marked with an asterisk in the table at § 712.30.

price their products. Thus, when manufacturers report about their own activities, the best estimate from readily obtainable data would be sufficiently accurate. There is no need for the company to expend further effort to report more exactly." 47 Fed. Reg. 26,994 (1982).

Reportable information on plant activities includes: (1) data pertaining to the amount of the chemical manufactured or imported; (2) the amount routinely lost during manufacturing operations; (3) a description of the amount processed in enclosed, controlled release, or open process operations; and (4) the amounts and number of workers involved for each. The level of accuracy for losses of the chemical during manufacturing must be specified by the company. EPA decided that companies should specify the level of accuracy in response to comments that the lack of measured data on such losses prevented respondents from meeting the "best estimates from readily obtainable data" standard. In determining the number of workers involved, the total number of persons employed directly in manufacturing, processing, and handling the chemical during the reporting period is to be counted.

A worker is to be considered as working with only one process category. If someone works in several process categories, the category in which the worker spends the most time is the one for which that worker should be counted.

Quantity in Commercial Distribution

The quantity of the substance prepared for distribution for industrial and consumer products is to be reported. "Industrial" use means the manufacturing and service industries included in the Standard Industrial Classification (SIC) codes, while consumer products are those used primarily by the general population. Industrial and consumer products are divided into three types: (1) chemical substances or mixtures; (2) articles or products with no release; and (3) articles or products with some release. In addition, the quantity of products for export, the quantity of chemicals consumed as a reactant, and the amount of unknown customer use are to be reported. This information must be accurate to within $\pm$ 50 percent. If the respondent cannot meet this accuracy level, the respondent should indicate on the form that the accuracy level is "unknown."

The Comprehensive Assessment Information Rule (CAIR) — Overview

EPA published the CAIR to provide a single form that can be used by the Agency under a variety of situations to elicit information on chemical substances under TSCA § 8(a). 53 Fed. Reg. 51,698 (1988). Nineteen substances were included on the initial CAIR list. EPA will continue to use this "model rule" and its accompanying form in the future as it identifies needs for information on additional chemicals by adding chemicals to the list of those already covered by the CAIR.

The CAIR differs from the PAIR in three major respects. First, only manufacturers and importers are subject to the PAIR, whereas processors, as well, are subject to the CAIR. Second, the CAIR reporting form is longer and more detailed than the PAIR form. PAIR respondents must answer all questions on the PAIR reporting form, but CAIR respondents are required to answer only selected questions. Using the basic 141-page CAIR reporting form, EPA can tailor data requests to meet the Agency's specific needs for particular information on a chemical by requesting answers to appropriate questions. Third, EPA presently can amend the list of substances subject to the CAIR only by notice-and-comment rulemaking. While additions to the PAIR list also are possible through notice-and-comment rulemaking, most substances on the PAIR list are added automatically when the Interagency Testing Committee designates or recommends that EPA require testing under TSCA § 4. The Agency intends eventually to amend the CAIR to allow this same automatic listing provided under the PAIR. At that time EPA intends to withdraw the PAIR.

The CAIR is designed to reduce duplicative reporting, conserve EPA and industry resources, and reduce reporting errors. EPA estimates that it will take one-quarter to one-third less work to add a substance to the CAIR list than it does to pass a chemical-specific rule. The CAIR's standardized reporting form also should facilitate data processing and review within EPA, and enable EPA to compile a single computerized data base to store and retrieve significant information. EPA views the CAIR as an important part of the Agency's overall long-term solution to data gathering.

Who Must Report

Under the CAIR, companies that manufacture, import <u>or process</u> a listed chemical substance are potentially subject to CAIR reporting requirements.[5/] 53 Fed. Reg. 51,718 (1988) (to be codified at 40 C.F.R. § 704.206(a)). If the CAIR requires reporting of processing activities for a chemical, a respondent which both manufactures and processes the chemical need report only on its processing activities.

Reporting requirements for manufacturers, importers and processors may differ. For example, EPA might require processors of a particular chemical to answer the CAIR form questions 10.08 through 10.16, and require importers and manufacturers of that chemical to answer questions 10.09 through 10.16. It is critical, therefore, to understand the distinctions between processing and manufacturing. This distinction must be drawn for each plant site, and the respondent must submit a separate CAIR form for each CAIR-listed chemical at each plant site.

Manufacturing. Manufacturing includes all activities at a site that are necessary to produce the listed substance and to make it ready for use or sale as the listed substance. 53 Fed. Reg. 51,718 (1988) (to be codified at 40 C.F.R. § 704.203). This includes processes where a listed substance is produced incidentally during the production of another substance. 53 Fed. Reg. 51,716 (1988) (to be codified at 40 C.F.R. § 704.3). Under TSCA § 3(7), importers are manufacturers. As a result, importers of a listed substance must report whenever the CAIR requires manufacturers to do so, but the converse is not necessarily true. The CAIR may require importers, but not manufacturers, to report on a particular substance.

As under PAIR, manufacturers of a chemical extracted from ores or other natural materials should report on the activities required to produce a listed chemical from natural source material, but not on the actual production of the natural source material.

Importing. An importer is any person who imports a chemical substance or imports a chemical substance as part of a mixture or article into the Customs territory of the United States. This includes the person primarily liable for payment of duties or that person's authorized agent. 53 Fed. Reg. 51,715 (1988) (to be codified at 40 C.F.R. § 704.3). It also includes the consignee, the importer of record, the actual owner (if an actual owner's designation and superseding bond has been filed in accordance with 19 C.F.R. § 141.20), and the transferee, if the right to draw merchandise in a bonded warehouse has been transferred in accordance with 19 C.F.R. pt. 144, subpt. C. <u>Id</u>.

Processing. Processing is preparation of a chemical substance or mixture for distribution in commerce. 53 Fed. Reg. 51,716 (1988) (to be codified at 40 C.F.R. § 704.3). "Processing" includes: (1) mixing or reacting the chemical with other chemicals or preparing it after its manufacture to make <u>another</u> substance for sale or use; (2) repackaging a listed

---

[5/] Manufacturers, importers and processors are only <u>potentially</u> subject to reporting, because some activities that are usually considered manufacturing or processing are specifically exempted from the CAIR. Exemptions are discussed in a separate section below.

chemical; and (3) purchasing a listed substance and then preparing it for use or distribution in commerce. 53 Fed. Reg. 51,718 (to be codified at 40 C.F.R. § 704.203).[6]

EPA has released two CAIR question and answer (Q&A) documents to clarify the often blurred distinction between manufacturing and processing. The first of these question and answer documents was released in December of 1988. In the latest question and answer document, dated March 1989, the Agency notes:

> To qualify as a manufacturer, a person must gain a commercial benefit from the production of the CAIR listed substance, but need not distribute the CAIR listed substance in commerce. A processor, however, must distribute in commerce either the CAIR listed substance or another substance produced using the CAIR listed substance as a reactant (a chemical producer). Therefore, if a company "uses" 100% of his processing capacity on-site as a final product or as an end-use product, that person is not a processor. "Use" of a CAIR listed substance as a reactant or as a chemical intermediate to manufacture a completely different substance which is distributed in commerce, however, is considered processing of a CAIR listed substance (chemical producer).

Q&A document at 1.

The documents indicate that manufacturers include companies that use raw non-CAIR materials to produce a CAIR-listed substance, import a CAIR-listed substance into the United States, or purify low grade material containing a CAIR-listed substance to produce a higher grade CAIR-listed substance. According to the Q&A document, a company that produces a CAIR-listed substance and adds additives or stabilizers before distributing the final product does not thereby become a processor of the listed substance. A company is a processor, however, if it buys the CAIR-listed substance and adds additives or stabilizers before distribution. If a company buys one CAIR-listed substance A and uses it to make a second CAIR-listed substance B, it processes A and manufactures B.

Notification of Trade Name Customers

Processors who buy a CAIR-listed substance under a trade name for processing may be unaware that it is listed. Therefore, the CAIR requires that "X/P" companies (manufacturers and processors who sell the CAIR-listed substances to processors under trade names) either must report on behalf of these customers or must notify them of their reporting obligation.

---

[6] Definitions of terms used in the CAIR are found in the three following separate sources: the statute; the CAIR regulations; and the CAIR Reporting Form Instructions Glossary (EPA, General Instructions -- EPA Form 7710-52, Comprehensive Assessment Information Rule Reporting Form). Thus, as an example, to understand how the CAIR defines "processor," one must look at the regulations and the glossary. The CAIR regulations define "processing activities" to include repackaging and define "repackager" as a person who buys a substance or mixture, removes it from the container in which it was bought and transfers it to another container. 53 Fed. Reg. 51,718 (1988) (to be codified at 40 C.F.R. § 704.203). According to the glossary, however, a container is "[a]ny free-standing, portable device in which substances are stored, treated, or otherwise handled." Thus a company that purchases and empties a 55 gallon drum of a CAIR-listed substance into a tank that is bolted to the ground is not subject to the CAIR as a processor solely on the basis of that action because the transfer was not to a "container" as defined in the glossary.

In meeting this requirement, X/P manufacturers and processors are given three options: 1) submit to EPA (no later than one day after the effective date of the Federal Register notice announcing the requirement) a list of all trade names under which they distribute the CAIR-listed substance; 2) report for their customers directly to EPA; or 3) notify customers of their reporting obligation by certified mail, return receipt requested.[7] Those who choose this last option must retain the return receipt for Agency inspection. They also must notify their customers, regardless of whether the customer already knows of its reporting obligation and regardless of whether their customer is exempt from CAIR reporting.

Temporary Administrative Relief For X/P Companies

On April 10, 1989, EPA issued a notice of temporary administrative relief from § 704.208(a) of the CAIR in response to a petition from the Synthetic Organic Chemical Manufacturers Association (SOCMA). 54 Fed. Reg. 14,324. SOCMA had expressed concern that compliance with this provision could result in disclosure of trade secret information.

Under the provisions of the notice of relief, any manufacturer, importer, or processor of a substance designated "X/P" in the CAIR list must comply with § 704.208(a) unless that party believes that it is unable to report for its customer(s), and that providing EPA with the trade names under which it distributes the CAIR-listed chemical or notifying its processor-customer would result, directly or indirectly, in the disclosure of a protected trade secret concerning the substance. Any party that has already submitted a trade name to EPA in compliance with 53 Fed. Reg. 51,719 (1988) (to be codified at 40 C.F.R. § 704.208(a)(1)) may still claim this temporary relief by notifying EPA that it believes that EPA's publication of the trade name would disclose a trade secret concerning the substance.

Any party which provides or has provided the specific identity of a CAIR substance in a trade name product to its customers through a material safety data sheet, an OSHA hazard communication, or some other mechanism, must comply with § 704.208(a).

Parties claiming this relief must notify EPA in writing and explain their basis for relief. This notification must have been postmarked no later than May 10, 1989. This relief is available only until the Agency has reconsidered § 704.208(a). EPA will reconsider § 704.208(a) when it proposes to add further "X/P" substances to the CAIR list.

Exemptions From The CAIR Reporting Requirements

In the CAIR, EPA has expanded the regulatory exemptions to the TSCA § 8(a) reporting rules. 53 Fed. Reg. 51,717 (1988) (to be codified at 40 C.F.R. § 704.5). These new provisions exempt companies that process or import a CAIR-listed substance solely as part of an article, manufacture a CAIR-listed chemical solely as a non-isolated intermediate, or manufacture, import or process small quantities of the substance solely for R&D.

---

[7] A customer is any person to whom a manufacturer, importer or processor directly distributes any quantity of a chemical substance, mixture, or article containing the substance or mixture, whether or not a sale is involved. 53 Fed. Reg. 51,698, 51,715 (1988) (to be codified at 40 C.F.R. § 740.3). Commentators criticized this definition as being too broad because it would require disposal facilities to report under the CAIR. The Agency, however, believes that there is a significant potential for exposure at disposal facilities and therefore intentionally included them in reporting requirements.

Additionally, the CAIR exempts small manufacturers, importers and processors.[8/] 53 Fed. Reg. 51,717 (1988) (to be codified at 40 C.F.R. § 704.5) and 53 Fed. Reg. 51,720 (1988) (to be codified at 40 C.F.R. § 704.210). Companies that only repackage the CAIR-listed substances are exempt as well, if they do not engage in any other processing, and are not otherwise covered. 53 Fed. Reg. 51,720 (1988) (to be codified at 40 C.F.R. § 704.210). Finally, companies which previously reported information voluntarily to EPA or another federal agency, using a CAIR form, may be partially exempt from current requirements. 53 Fed. Reg. 51,720 (1988) (to be codified at 40 C.F.R. § 704.210(c)).

Reporting Period

The CAIR specifies the time period during which the reportable activity must have taken place in order to trigger reporting requirements. In determining this period the respondent should follow a three step process: (1) determine which, if any, corporate fiscal years fall completely within any CAIR-designated coverage period; (2) determine whether it engaged in an activity subject to that CAIR listing during any of those fiscal years; and (3) report only activities that took place during the most recent corporate fiscal year falling completely within a CAIR designated coverage period. A company need not report if it did not manufacture, import or process a CAIR-listed chemical. See 53 Fed. Reg. 51,720 (1988) (to be codified at 40 C.F.R. § 704.214).

Chemicals Subject To Reporting

Two EPA program offices (the Office of Air and Radiation (OAR) and the Office of Toxic Substances (OTS)) and three other federal agencies nominated forty-seven chemical substances and mixtures for the proposed rule. The final rule lists only the following nineteen, which were nominated by OAR, OTS and the National Institute for Occupational Safety and Health (NIOSH):

	CAS No.
Hydrazinecarboxamide	57-56-7
Acetamide (Ethanamide)	60-35-5
Ethane, 1,1,2,2-tetrabromo-	79-27-6
Phenanthrene	85-01-8
Benzene, 1,3-diisocyanato-2-methyl- (2,6-Toluene diisocyanate)	91-08-7
2-Naphthalenesulfonic acid, 6-amino- (Broenner's acid)	93-00-5
Benzenamine, 4,4'-methylenebis[2-	101-14-4

[8/] The CAIR definition of "small manufacturer" differs from the PAIR definition, and is applied to importers and processors as well as to manufacturers. Thus, a company that manufactures, imports or processes less than 100,000 lbs. (45,400 kg.) of the listed chemical annually and has total annual sales of less than $40 million is exempt. A manufacturer, importer or processor with total annual sales of less than $4 million is exempt regardless of the amount of the listed chemical produced or processed. These annual sales figures include those of any parent company.

chloro- (MBOCA)

Ethanol, 2-chloro-, phosphate (3:1) (Tris(2-chloroethyl) phosphate)	115-96-8
Pyrene (Benzo[def]phenanthrene)	129-00-0
Hydrazinecarboxamide, monohydrochloride (Semicarbazide hydrochloride)	563-41-7
Benzene, 2,4-diisocyanato-1-methyl- (2,4-Toluene diisocyanate)	584-84-9
Disulfide, dimethyl (Dimethyl disulfide)	624-92-0
Benzene, diisocyanatomethyl- (Unspecific toluene diisocyanate)	1321-38-6
Hydroxylamine, hydrochloride (Hydroxylammonium chloride)	5470-11-1
Chlorine	7782-50-5
Hydroxylamine (Oxammonium)	7803-49-8
Hydroxylamine, sulfate (2:1) (Hydroxylammonium)	10039-54-0
Hydroxylamine, sulfate (1:1) (Hydroxylamine acid sulfate)	10046-00-1
Benzene, 1,3-diisocyanatomethyl- (Toluene diisocyanate)	26471-62-5

53 Fed. Reg. 51,722 (1988) (to be codified at 40 C.F.R. § 704.225).

These final nineteen chemicals were selected for one of three reasons: (1) the agencies lack current exposure data and know or suspect the chemical causes adverse health or environmental effects; (2) the agencies believe data gaps for the chemical are significant; or (3) the agencies place a priority on the need for the requested information to complete assessments of the substances. 53 Fed. Reg. 51,698, 51,706 (1988).

Specific Information Requirements

Subpart D of the CAIR regulations sets out a matrix that identifies specific reporting requirements for each listed chemical. All respondents must answer all questions in part A of the CAIR form. The subpart D matrix specifies which additional questions respondents must answer. One CAIR form must be submitted for each site at which a listed chemical is manufactured, imported, or processed for each activity. If a company is engaged in more than one activity with a listed chemical at a single site, they must report on all specified activities on the same form. 53 Fed. Reg. 51,720 (to be codified at 40 C.F.R. § 704.212(c)). In completing the CAIR form, a respondent must report information that is "known to or reasonably ascertainable by" it. 53 Fed. Reg. 51,704 (1988). See also 40 C.F.R. § 704.3.

The CAIR form elicits ten different classes of information:

A. General Manufacturer, Importer, and Processor Information.

B. Manufacturer, Importer, and Processor Volume and Use.

C. Processor Raw Material Identification.

D. Physical/Chemical Properties.

E. Environmental Fate.

F. Economic and Financial Information.

G. Manufacturing and Processing Information.

H. Residual Treatment, Generation, Characterization, Transportation, and Management.

I. Worker Exposure.

J. Environmental Release.

EPA, Comprehensive Assessment Information Rule Reporting Form (EPA Form 7710-52) (n.d.) (CAIR form).

The CAIR form instructions set out detailed directions for responding to questions in each of these areas. EPA, General Instructions [for] EPA Form 7710-52, [CAIR] Reporting Form (n.d.). The following descriptions summarize the types of information required under each of these ten categories.

Recordkeeping Requirements

For a period of three years, respondents must maintain a copy of the CAIR form submitted to the Agency, supporting materials sufficient to verify or reconstruct the report, and a copy of all notices sent to and return receipt cards received from customers of those who distribute a listed substance under a trade name. 53 Fed. Reg. 51,717 (1988) (to be codified at 40 C.F.R. § 704.11).

Confidential Business Information

Respondents must assert and substantiate confidential business information (CBI) claims at the time the CAIR form is submitted, or EPA will consider the right to make such claims waived. Respondents who assert that any of the submitted information is CBI must submit both their full CAIR form and a second "sanitized" copy from which all the purported CBI has been deleted. The CAIR sets forth the process by which a respondent may claim information as confidential. 53 Fed. Reg. 51,721 (1988) (to be codified at 40 C.F.R. § 704.219). Appendix II of the CAIR form contains a CBI substantiation form which must accompany the CAIR form that contains the purported CBI. All information not claimed and substantiated as confidential will be placed in a public file.

Coordination with Other EPA Reporting Requirements

While the CAIR was not promulgated to replace all other information-gathering authority, the Agency intends to coordinate reporting requirements under other acts with those of the CAIR rule and to use information submitted on the CAIR form where possible to meet other statutory obligations. Information gathered under the CAIR will be shared among EPA offices to the extent possible, eliminating much duplicative reporting. For example, the CAIR does not ask respondents to answer questions that directly overlap questions already asked under the SARA § 313 rule. 53 Fed. Reg. 51,699 (1988). Likewise, the Agency will not ask for information similar to that already collected on a substance under the Agency Accidental Release Information Program. Id. Finally, other agencies or EPA offices will continue to use their own authority when the substance of interest cannot be regulated under TSCA, the information sought could be gathered more efficiently under another authority, or the information sought is more specific in detail than would be elicited on the CAIR reporting form.

MAINTAINING THE TSCA INVENTORY

From 1975 through 1979 EPA required manufacturers and importers of all chemicals to report information on identity, production or import volume, and whether use was site-limited. 40 C.F.R. § 710.3. Based on this initial reporting, EPA established the TSCA Inventory. In order to keep the information on the Inventory current, the Agency has also established an ongoing reporting requirement.

TSCA § 8(b): Inventory Update Rule

In June 1986, EPA issued an Inventory Update Rule requiring manufacturers and importers of certain chemicals listed on the Inventory to report current data on the production volume, plant site and site-limited status of the substances. 51 Fed. Reg. 21,438 (1986) (codified at 40 C.F.R. § 710, subpt. B). The Agency also made available detailed instructions for manufacturers and importers subject to the Update Rule. See EPA, Office of Toxic Substances, Information Management Division, Instructions for Reporting for the Partial Updating of the TSCA Chemical Inventory Data Base (April 1986), available from the TSCA Assistance Office. The initial 120-day reporting period commenced on August 25, 1986 and ended on December 23, 1986. Recurring reporting is required every four years. Four categories of substances listed on the Inventory are exempt from the Update Rule's reporting requirements: (1) polymers; (2) microorganisms; (3) naturally occurring substances; and (4) inorganics.[9/] 40 C.F.R. § 710.26.

Any company which manufactures or imports for commercial purposes 10,000 pounds or more of a "reportable substance" at any time during the most recent complete corporate fiscal year immediately preceding the reporting period is obligated to submit updated information. 40 C.F.R. § 710.28. Small manufacturers, however, are exempt from reporting. 40 C.F.R. § 710.29. In addition, companies which manufacture small quantities of reportable substances solely for R&D, import the substance as part of an article, or manufacture the substance as an impurity by-product, or in a manner incidental to its end use, are exempt. 40 C.F.R. § 710.30. This small manufacturer exemption, however, does not apply with regard to the manufacture of any substance that is the subject of: (1) a proposed or final rule under TSCA §§ 4, 5(b)(4), or 6; (2) an order under TSCA § 5(e); or (3) relief granted pursuant to a civil action under §§ 5 or 7 of TSCA. 40 C.F.R. § 710.29.

[9/] Many polymers, naturally occurring substances, and inorganic chemicals are "flagged" in the 1985 edition of the Inventory to indicate that they are not subject to the Update Rule.

Companies subject to the Update Rule must report the specific chemical identity, the plant site, whether the substance is manufactured or imported, whether the substance is distributed for commercial purposes off the site of manufacture or import, and the production volume in pounds. 40 C.F.R. § 710.32. This information can either be reported by completing an original numbered reporting form available from EPA or by using a computer tape. The Update Rule also requires each manufacturer or importer subject to the rule to maintain specific records documenting the information submitted to EPA. 40 C.F.R. § 710.37. Importantly, for those substances manufactured at less than 10,000 pounds, production volume records must be maintained to justify a decision not to report.

Special Reporting Problems

Although seemingly straightforward, the Update Rule creates a number of special reporting problems. One such problem arises when a company divests a division or subsidiary which had manufactured or imported reportable chemical substances. The question of which corporate entity bears the responsibility for reporting depends upon what was transferred during divestiture. If an entire division was sold (with all of its operations, records, assets, obligations, and liabilities) prior to the end of a reporting period, the transferee would be required to report. The transferor would be obligated to report, however, if the transfer occurred any time after the end of the reporting period. Letter from Linda A. Travers, Director, Information Management Division, to McKenna, Conner & Cuneo (Aug. 15, 1986).

Other questions pertain to the parent company-subsidiary relationship and the small manufacturer exemption. EPA has indicated that the reporting obligation rests with the company subject to the rules even if it is a subsidiary of a larger company. 51 Fed. Reg. 21,445 (1986). EPA also has stated that the definition of "small manufacturer" includes the total annual sales of parent firms and subsidiaries. This is because "EPA believes that a broadly defined sales criterion is the best available measure of a diversified firm's full financial resources available for regulatory compliance." Id.

TSCA § 8(c): RECORDS OF SIGNIFICANT ADVERSE REACTIONS

TSCA § 8(c) requires manufacturers, processors, and distributors to keep records of significant adverse reactions to health and the environment alleged to have been caused by a chemical substance or mixture. Such allegations may be made by "any source," including employees, customers, surrounding neighbors, or companies on behalf of their employees, or organizations on behalf of their members. Employee allegations must be kept on file for thirty years; allegations by others for five years. These allegations do not have to be reported to EPA unless the Agency specifically requests them.

EPA's rule implementing TSCA § 8(c) establishes a recordkeeping and reporting system that will "[c]reate a historical record of significant adverse reactions alleged to have been caused by [a particular] substance or mixture" and to "[p]rovide a means to identify previously unknown chemical hazards and to reveal patterns of adverse effects" which previously might have gone undetected. 48 Fed. Reg. 38,178 (1983) (supplementary information to final rule (codified at 40 C.F.R. pt. 717)). The TSCA § 8(c) rule became effective November 21, 1983.

Persons Subject to the Rule

Manufacturers of chemical substances and mixtures covered by TSCA (with the exception of manufacturers involved solely in mining or other extractive functions -- mineral, oil, petroleum, natural gas, coal, etc.) are subject to the provisions of 40 C.F.R. pt. 717. The manufacture of a chemical substance for commercial purposes at any site owned or controlled by a firm is sufficient to subject that corporate entity to the § 8(c) rule. If the United States manufacturing facility is the subsidiary of a foreign parent, the subsidiary is the "person" subject to § 8(c). United States parents with foreign subsidiaries, however, are responsible for recording allegations originating with the foreign subsidiary. The degree of United States parent involvement with the foreign subsidiary is irrelevant. Although federal agencies, such as the Department of Defense, do not engage in commercial activities and, therefore, are not manufacturers for a commercial purpose, government-owned, company-operated facilities are considered manufacturers within the meaning of the Act.

The original rule made processors in Standard Industrial Classification (SIC) code 28 (chemicals and allied products) and SIC code 2911 (petroleum refining) subject to § 8(c). EPA later abandoned the use of SIC codes to identify processors who are subject to § 8(c). 50 Fed. Reg. 46,769 (1985). As amended, the rule exempts all processors except those who process chemical substances to produce mixtures, or who repackage chemical substances or mixtures. 40 C.F.R. § 717.5(b). In other words, formulators and repackagers are subject to § 8(c). EPA also has exempted from the rule's provisions sole distributors, defined as a person "solely engaged in the distribution of chemical substances," and retailers. There are no exemptions for small manufacturers and processors. 40 C.F.R. § 717.7(c), (d).

What Constitutes a Significant Adverse Reaction

EPA has defined a significant adverse reaction to mean "reactions that may indicate a substantial impairment of normal activities, or long-lasting or irreversible damage to health or the environment." 40 C.F.R. § 717.3(i).

Human Health Reactions

This definition is intended to encompass not only damage that manifests itself physiologically, such as cancer or birth defects, but also includes "subjectively experienced adverse reactions" which, although not physically verifiable, may be indicative of a more serious effect. EPA intends that recordkeepers file those reactions which initially appear less serious in nature, but could indicate a pattern of adverse effects. The Agency concedes that recordkeepers must exercise a considerable amount of judgment in determining which adverse reactions are significant ones.

EPA has listed four groups of reactions at 40 C.F.R. § 717.12 which are intended to provide examples of the range of reactions that should be considered "significant." These types of reactions include: (1) long-lasting or irreversible damage, such as cancer or birth defects; (2) partial or complete impairment of bodily functions, such as reproductive disorders, neurological disorders or blood disorders; (3) an impairment of normal activities experienced by all or most of the persons exposed at one time; and (4) an impairment of normal activities which is experienced each time an individual is exposed. 40 C.F.R. § 717.12.

Reactions covered by items three and four encompass two criteria. First, they involve the impairment of a normal activity (both job and non-job related). Second, they are repeated, either by a group of persons, or by a single individual several times.

In order to place some limitation on what initially appears to be an open-ended recording obligation, EPA has provided a narrow exemption for known human health effects. 40 C.F.R. § 717.12(b). Known human health effects are defined as "commonly recognized human health effect[s] of a particular substance or mixture" as described in scientific articles and publications, or the firm's material safety data sheets. 40 C.F.R. § 717.3(c)(1). An effect is not considered to be a known human health effect, however, if the effect: was a significantly more severe toxic effect than previously described; was a manifestation of a toxic effect after a significantly shorter exposure period or lower exposure level than described; or was a manifestation of a toxic effect by an exposure route different from that described. 40 C.F.R. § 717.3(c)(2).

EPA has stated that the results of in vitro or animal testing on the substance cannot be considered the equivalent of known human health effects. Thus, an allegation that a chemical produced a chronic or subchronic effect in a human must be recorded if, at the time of the allegation, animal studies are the only evidence that the chemical can produce that effect. The Agency has qualified this exemption of adverse effects in animals. In a question and answer document issued on the § 8(c) rule, the Agency stated that acute effects (e.g., primary irritant effects) demonstrated only in animal test species "may be considered known human effects because of the inevitability that such substance will have similar adverse acute effects on human tissue." EPA, Office of Toxic Substances, Questions and Answers Concerning the TSCA Section 8(c) Rule at 45 (July 1984) (8(c) Q&A July 1984).

Environmental Reactions

Environmental reactions which must be recorded include: (1) gradual or sudden changes in the composition of animal life or plant life, including fungal or microbial organisms, in an area; (2) abnormal number of deaths of organisms (e.g., fish kills); (3) reduction of the reproductive success or the vigor of a species; (4) reduction in agricultural productivity, whether crops or livestock; (5) alterations in the behavior or distribution of a species; and (6) long lasting or irreversible contamination of components of the physical environment, especially in the case of ground water, and surface water and soil resources that have limited self-cleansing capability. 40 C.F.R. § 717.12(c). These reactions are to be recorded even if the incident is restricted to a single plant or disposal site.

EPA has provided a limited exemption from the requirement to record environmental reactions. If the alleged cause of the significant adverse reaction can be "directly attributable" to an accidental spill or discharge, an emission exceeding permitted levels, or other incidents which have been reported to the federal government under any applicable authority, firms are not required to record the incident under § 8(c). 40 C.F.R. § 717.12(d). For example, releases at a Superfund site are not recordable if the criteria at 40 C.F.R. § 717.12(d) are met. Environmental contamination incidents reported to local or state authorities, however, are not exempt unless such government has been delegated responsibility for administering the federal law in question. EPA developed this exemption primarily to avoid the problem of duplicative recordkeeping and reporting of environmental incidents. The Agency declined to base the exemption on "known environmental effects" because of the wider variability involved in environmental reactions than in human reactions.

Allegations Which Must Be Recorded

An "allegation" is defined as "a statement, made without formal proof or regard for evidence, that a chemical substance or mixture has caused a significant adverse reaction to health or the environment." 40 C.F.R. § 717.3(a) (emphasis added).

In order to constitute an allegation which is recordable under § 8(c), the statement must clearly state the alleged cause of the adverse reaction by identifying one or more of the following: (1) a specific substance; (2) a mixture or article that contains a specific substance; (3) a company process or operation in which substances are involved; and (4) an effluent, emission, or other discharge from a site of manufacturing, processing or distribution of a substance. 40 C.F.R. § 717.10(b)(2). Under this definition, no proof or evidence, such as a doctor's report, is required for an allegation to be recordable.

Allegations can be derived from a variety of sources. For example, a lawsuit or other legal action which meets the criteria of § 8(c) can constitute an allegation. In addition, discovery papers can constitute part of the allegations. Companies have raised questions about the recordability of allegations received by "off-duty employees at social functions." According to the Agency, allegations only require action under § 8(c) when they have been received by a person subject to the rule's provisions. In general, allegations made in such a situation would not be "received" within the meaning of § 8(c). EPA has cautioned, however, that if the off-duty employee is an official who has knowledge of the company's § 8(c) responsibilities, that official should request that the alleger submit the allegation in writing or otherwise contact the official during normal business hours. Each allegation must be recorded separately, even if several are identical or are very similar allegations received over a short period of time. It is important to remember that a series of identical or very similar allegations about a particular substance may indicate a significant risk, which can trigger reporting requirements under § 8(e) (see below for a discussion of § 8(e)).

Both written and oral allegations are subject to the § 8(c) rule. An oral allegation can be handled in one of two ways. A company may transcribe the allegation for recordkeeping purposes. There is no requirement that company transcriptions be signed by the alleger. Alternatively, the company can request that the alleger submit the allegation in writing and sign it. 40 C.F.R. § 717.10(b)(1). (The company request does not have to be in writing; an oral request is sufficient.) If the alleger fails to respond to the company's request for a written allegation, the company is not required to take any further action with respect to § 8(c). For example, companies are not required to record unsigned union grievances. EPA has provided a number of additional hypothetical examples to illustrate the rule's provisions in two question and answer documents. These documents are part of the official public rulemaking record and may be available from the EPA TSCA Assistance Office. EPA, Office of Toxic Substances, TSCA Section 8(c) Final Rule Questions and Answers (Nov. 1983) (8(c) Q&A Nov. 1983) and 8(c) Q&A July 1984.

The § 8(c) rule does not designate which company official shall be responsible for recognizing and recording oral or written allegations. Rather, companies are free to designate the responsible official. Companies must educate their employees, however, as to the identity of the official.

Recordkeeping Requirements

Only those allegations meeting the criteria of 40 C.F.R. § 717.10(b)(2) are required to be recorded. EPA does not require allegations to be recorded on a specific form. In general, firms are permitted to keep the information in a format of their own choosing as long as the original allegation is retained and an abstract of the allegation containing the following information is prepared and recorded:

(i) The name and address of the plant site which received the allegation.

(ii) The date the allegation was received at that site.

(iii) The implicated substance, mixture, article, company process or operation, or site discharge.

(iv) A description of the alleger (e.g., "company employee," "individual consumer," "plant neighbor"). If the allegation involves a health effect, the sex and year of birth of the individual should be recorded, if ascertainable.

(v) A description of the alleged health effect(s). The description must relate how the effect(s) became known and the route of exposure, if explained in the allegation.

(vi) A description of the nature of the alleged environmental effect(s), identifying the affected plant and/or animal species, or contaminated portion of the physical environment.

40 C.F.R. § 717.15(b).

Although a company can utilize a computer data base to manage its § 8(c) records, EPA requires that the original allegation or a microfiche copy be retained; a computer data base is not viewed as a suitable replacement for a hard copy. 8(c) Q&A July 1984 at 16. Firms are not required to investigate the allegations, but must record the results of any investigation conducted. Finally, copies of any required records or reports relating to the allegations must be kept. For example, if the allegation must be reported to OSHA, a copy of that OSHA report must be maintained in the file. An allegation must be recorded based upon its original contents, even if some aspect is incorrect. Once recorded and filed, such an allegation may not be removed even if a subsequent investigation reveals inaccuracies in the allegation. Rather, the results of any further investigation may be placed in the allegation file.

Records of allegations of employees must be maintained in a central location for thirty years from the date the reactions were first reported or known to the person maintaining the records. All other records of significant adverse impact must be maintained for a period of five years. The unintentional loss or destruction of a firm's § 8(c) files will be handled by the Office of Compliance Monitoring on a case-by-case basis. If a firm ceases to do business, the successor firm must receive and maintain the § 8(c) records. If there is no successor firm, the records must be forwarded to EPA. 40 C.F.R. §§ 717.15(d), (e)

These § 8(c) records are subject to EPA inspection. In conducting an inspection, EPA will examine whether the company possesses a basic awareness of its obligations under § 8(c). Such a basic awareness is demonstrated by the presence of a system designed to forward allegations to a central recordkeeping location. EPA also will inspect the actual records to determine if all the information requirements previously described are present, and whether allegations, when taken together, may reasonably support a conclusion of substantial risk under § 8(e) (see discussion below).

EPA can require submission of copies of the § 8(c) records to the Agency upon request and employees can petition the Agency to collect and release § 8(c) information. In 1986, EPA required two manufacturers to submit § 8(c) records on allegations relating to perfluoroalkyl resins. EPA's request was sent by letter to the two companies, not announced in a Federal Register notice. In 1988, EPA issued a Federal Register notice that called in § 8(c) records from manufacturers, importers, processors and even certain distributors of tri (alkyl/alkoxy) phosphates and diisocyanates. 53 Fed. Reg. 1408 (1988).

SECTION 8(d): HEALTH AND SAFETY STUDIES

Section 8(d) requires that, upon request, a person who manufactures, processes, or distributes in commerce any chemical substance or mixture, must submit to the Administrator lists and copies of health and safety studies conducted by, known to, or ascertainable by that person. Thus, § 8(d) is a potentially broad reporting provision because it mandates the reporting of all pertinent studies on a chemical.

Information from these studies is intended to be used in making regulatory decisions under TSCA §§ 4, 5, and 6. For example, EPA may obtain exposure information on a chemical under § 8(a) and existing health and safety studies under § 8(d) to determine whether the Agency should promulgate a § 4 test rule to require manufacturers to conduct additional testing.

Model Reporting Rule

In 1982, EPA issued a final Model Health and Safety Data Reporting Rule to implement § 8(d). Model Reporting Rule, 47 Fed. Reg. 38,780 (1982) (codified at 40 C.F.R. pt. 716). On September 15, 1986, EPA published a final rule amending the Model Reporting Rule. 51 Fed. Reg. 32,720 (1986). Under the § 8(d) rule, submission of unpublished health and safety studies is required on certain specifically listed chemicals or mixtures. Chemicals designated or listed by the Interagency Testing Committee are automatically added to the list of chemicals. Other chemicals can be added to the list by notice-and-comment rulemaking.

EPA will require reporting of unpublished health and safety studies when the Agency is considering action to control exposure or to provide supporting data when EPA is conducting a risk assessment on a chemical.

Who Must Report

Persons who currently manufacture, import, or process a chemical substance or a mixture listed at 40 C.F.R. § 716.120 (or propose to do so) or who manufactured, imported or processed (or proposed to do so) within the ten years preceding the effective data of the listing of the chemical are subject to the provisions of the Model Reporting Rule. There are two phases to § 8(d) reporting. First, persons are required to submit copies of all non-exempt studies in their possession at the time they become subject to the rule. The copies must be submitted within sixty days after the addition of a chemical substance or mixture to the subject chemical list, and copies of any additional studies that are in progress must be submitted within thirty days of their completion. In addition, EPA must be informed within thirty days of any study on a subject chemical <u>initiated</u> by or for such manufacturer or processor after the initial sixty-day reporting period. This requirement continues for a period of ten years and applies to manufacturers or processors who begin to manufacture or process a listed chemical or designated mixture any time prior to the expiration of the ten-year period. 40 C.F.R. §§ 716.60, 716.65.

Section 8(d) is retroactive in its application; anyone who has manufactured, imported, or processed a listed chemical or designated mixture anytime in the preceding ten years also must submit copies of studies in their possession within sixty days after the chemical is added to the subject list, even if the person no longer manufactures or processes the chemical. 40 C.F.R. §§ 716.5(a)(1), 716.60.

Health and Safety Study

EPA defines the term "health and safety study" to mean "any study of any effect of a chemical substance or mixture on health or the environment or on both, including underlying data and epidemiological studies, studies of occupational exposure to a chemical substance or mixture, toxicological, clinical, and ecological or other studies of a chemical substance or mixture, and any test performed under TSCA." 40 C.F.R. § 716.3.

EPA interprets this definition broadly so that other information relating to the effects of a chemical substance or mixture are included within the phrase "health and safety study." EPA has provided an extensive list of examples of the types of tests that constitute health and safety studies within the meaning of the Model Reporting Rule. 40 C.F.R.

§ 716.3. These examples include toxicity tests, tests for ecological and other environmental effects, assessments of human and environmental exposure, and monitoring data.

Monitoring Reports. Monitoring efforts that attempt to define exposure levels of manufacturing or processing workers must be submitted if the data are "aggregated and analyzed" and their meaning discussed in the study report. For example, if a company monitors the exposure of workers involved in the manufacture of a subject chemical and discusses the monitoring data in a report which analyzes the data to draw conclusions about occupational or environmental exposure, this report must be submitted. In contrast, daily or routine monitoring data, even if tabulated, do not have to be submitted if the report "merely confirms that permissible levels of a chemical have or have not been exceeded." EPA, Questions and Answers About Reporting Under the TSCA § 8(d) Health and Safety Study Reporting Rule at 1 (Nov. 10, 1982) (8(d) Q&A Nov. 1982).

In 1986, EPA brought its first, and to date only, civil action under TSCA § 8(d) against Lonza, Inc. seeking a $1.5 million penalty. Lonza, Inc., No. TSCA 87-H-03 (EPA, July 20, 1987). The key question raised by the case was whether "routine" monitoring of employee health for insurance purposes constituted a § 8(d) health and safety study. EPA eventually dropped the complaint against Lonza, Inc. after finding that the Agency had earlier taken the position (in writing) that very similar reports were not subject to TSCA § 8(d) reporting. In the wake of Lonza, EPA has recognized that there is considerable uncertainty regarding the reportability of monitoring studies. As a result, the Agency expects to release the final version of a new Question and Answer guidance document in 1989 to clarify when monitoring studies must be submitted under TSCA § 8(d).

Modeling Studies. The Model Reporting Rule considers modeling studies that estimate actual or reasonably likely environmental or human exposure to be reportable as "assessments of human and environmental exposure." By contrast, modeling studies based on very conservative or worst-case exposures need not be reported, but those that use a company's best estimates of emission quantities and conditions should be. For example, a modeling study conducted in planning plant or equipment construction does not have to be reported because it does not estimate actual or reasonably likely levels of existing environmental releases. As elsewhere under TSCA § 8(d), only studies and not data, computer printouts or hand-written calculations need be submitted unless specifically requested by the Agency under 40 C.F.R. § 716.10(a)(4). EPA, Office of Pesticides and Toxic Substances, Questions and Answers: Applicability of TSCA Section 8(d) Model Health and Safety Data Reporting Rule (40 C.F.R. pt. 176) to Modeling Studies (Dec. 29, 1988).

Company Operations Reports. Many companies routinely conduct surveys of their operations and report their findings in operations reports. These operations reports may contain portions discussing § 8(d) chemicals or mixtures. Only those portions which discuss § 8(d) chemicals or mixtures and which constitute a "study" need be reported -- not the entire report.

Clinical Tests on an Employee. An individual employee's clinical test results which appear in that employee's medical record should not be submitted. 8(d) Q&A Nov. 1982 at 4. Reports that characterize and discuss the implications of a chemical in the blood levels of workers, however, must be submitted since this constitutes a study of exposure in the work force, and the results clearly have implications beyond a single individual.

Reporting Physical/Chemical Properties. EPA's Model Reporting Rule requires the reporting of studies on any of ten physical/chemical properties if the studies are for the purposes of determining the environmental or biological fate of the substance or mixture. EPA uses this information to develop a picture of environmental exposure to a chemical and the chemical's effects in order to permit effective evaluation of potential risks. As a general guideline, environmental effects studies should be thought of as including any study of a chemical's effect on the environment, and monitoring data that has been aggregated and analyzed to measure the effect of the chemical on the environment. These physical chemical properties include: (1) water solubility; (2) adsorption/desorption on particulate surfaces, e.g., soil; (3) vapor pressure; (4) octanol/water partition coefficient; (5) density/relative density; (6) particle size distribution for insoluble solids; (7) dissociation constant; (8) degradation by photochemical mechanisms -- aquatic and atmospheric; (9) degradation by chemical mechanisms -- hydrolytic, reductive, and oxidative; and (10) degradation by biological mechanisms -- aerobic and anaerobic. 40 C.F.R. § 716.50.

Exemptions From Reporting

EPA has provided a number of exemptions from the reporting requirements of the Model Reporting Rule. 40 C.F.R. § 716.20. A discussion of exemptions follows.

1. <u>Studies Which Have Been Published in the Scientific Literature</u>. Any study which has been published in the scientific literature is exempt from reporting. According to EPA the term "scientific literature" is defined as "any periodical, book or monograph which presents data obtained through a systematic pursuit of knowledge involving the recognition and formulation of a problem, the collection of data through observation and experiment, and the formulation and testing of hypotheses." EPA, Office of Pesticides and Toxic Substances, General Comments on the Proposed Section 8(d) Rule; Support Document for the Final Section 8(d) Rule at 4 (Sept. 1, 1982).

2. <u>Studies Previously Submitted to OTS</u>. In 1986 EPA amended the exemption for studies previously submitted. 51 Fed. Reg. 32,720 (1986). As originally issued, studies previously submitted to any EPA office were exempt from reporting. Now, only studies previously submitted to OTS under TSCA § 4 proceedings, § 8(e) submissions, a premanufacture notice, or a significant new use rule, and studies submitted on a "for your information" basis are exempt from reporting. All other studies previously submitted to EPA with a claim of confidentiality are now subject to submission or listing.

3. <u>Studies Previously Submitted to a Federal Agency</u>. Studies submitted to EPA offices other than OTS or to other federal agencies with no claims of confidentiality are exempt only from the copy submission requirements. Lists of such studies must still be provided.

4. <u>Studies Conducted or Initiated By or For Another Person Subject to § 8(d)</u>. Studies conducted by or for another person on a chemical subject to the § 8(d) reporting rule must be reported. Many times studies are conducted by a parent for its subsidiary or vice versa. Companies are not required to acquire copies of studies from their foreign subsidiaries. Companies are required, however, to list the studies known to them, but not in their possession, if they know copies of the studies will not be submitted by the person who initiated or conducted the studies. In addition, EPA does not require companies to search for studies on chemicals manufactured by their foreign subsidiaries.

5. <u>Studies of Chemical Substances Which Are Not on the TSCA Inventory</u>. EPA has exempted chemical substances not on the TSCA Inventory from the reporting requirements of the § 8(d) model rule. This exemption applies only to substances within the categories listed at 40 C.F.R. § 716.120(c) (chemicals and mixtures subject to § 8(d)). Thus, all chemicals used solely for R&D are exempt from reporting. When a TSCA § 5 premanufacture notice is filed on that substance, however, any health and safety studies have to be submitted under 40 C.F.R. § 720.50.

6. _Certain Acute Toxicity Studies._ EPA has exempted from reporting five types of acute and primary toxicity studies commonly performed using mammalian species. This exemption applies only when the substance tested was a mixture which either contained a listed chemical substance or a listed mixture. Thus, acute oral toxicity, acute dermal toxicity, acute inhalation toxicity, primary eye irritation, and primary dermal irritation studies performed on mammals do not have to be submitted. Acute studies involving aquatic species, such as fish, however, must be reported. Similarly, mutagenicity tests on mixtures also must be submitted.

7. _Physical and Chemical Properties._ Only studies involving one of the ten physical/chemical characteristics listed at 40 C.F.R. § 716.20 need to be reported. Studies involving any other physical/chemical property are not subject to reporting.

8. _Analyzed Aggregations of Monitoring Data Acquired More than Five Years Ago._ Monitoring data collected more than five years before the substance or mixture was added to the § 8(d) list need not be reported, even if the data have been aggregated and analyzed.

9. _Impurities._ Studies of listed chemicals or mixtures manufactured or processed only as impurities are not subject to reporting requirements.

10. _Studies Previously Submitted by Trade Associations._ In the Agency's September 1986 amendments to its § 8(d) model rule, an express exemption was added for studies previously submitted by trade associations on behalf of their members. 51 Fed. Reg. 32,728. Trade associations must submit such studies within sixty days of the date on which the substance is listed. Thus, member companies whose trade association has submitted the required studies are exempt from both the copy and list submission requirements. In addition, EPA has specifically exempted studies on plant growth or damage from ureaformaldehyde resins when used as a fertilizer.

File Searches

Companies often design their filing systems to access a product, not the individual components of a product. In recognition of this, EPA has stated that companies can conduct an examination of non-exempt product studies (sub-chronic, chronic, environmental) by scanning the study and the appended formulation information to ascertain if any of the listed § 8(d) substances are identified. If no substances are identified and cannot be identified, no additional search is required.

Once a company determines that a chemical it manufactures or processes is subject to the § 8(d) reporting rule, it must search its files for health and safety studies on that substance. The model rule requires a reasonable search of files where the required information is ordinarily kept, including records kept by a company's employees responsible for health and environmental matters. In a Q&A document, EPA has stated that only sites in the United States must be searched. It remains unclear whether a foreign site must be searched if that is where a domestic company ordinarily keeps the required information.

Submission of Copies of Studies

Unless specifically exempted, persons who have manufactured or processed the listed substances or mixtures within the ten years preceding the effective date of the listing of the chemical must submit copies of health and safety studies in their possession on the substance or mixture. Underlying data, however, do not have to be submitted initially. Underlying data include data such as medical or health records, individual files, lab notebooks, and daily monitoring records.

Copies of required studies must be submitted with a cover letter which identifies the name, job title, address, and telephone number of the submitting office and the name and address of the manufacturing or processing establishment on whose behalf the submission was made. EPA has provided specific directions on how to report studies on listed substances and mixtures at 40 C.F.R § 716.30. If the substance is listed individually under 40 C.F.R. § 716.120(a), studies of the substance and studies of mixtures known to contain the substance must be reported as studies of that substance. When two or more substances are listed as a mixture under 40 C.F.R. § 716.120(b), the studies of the listed mixture (or mixtures known to contain the listed mixture) must be reported. 40 C.F.R. § 716.45(b). If the substance is an aqueous solution or if the substance contains a small amount of an additive, or varies in purity grade, the study must be reported as studies of the substance itself, not as a mixture. For example, if a listed chemical is added to a test substance (also listed) for the sole purpose of introducing the test substance into a testing system, the resulting studies would be reported as studies of the tested chemical substance, not as studies of the mixture formed for testing purposes. 40 C.F.R. § 716.45.

Submission of Lists of Studies

Unless specifically exempted, manufacturers and processors of listed chemical substances or mixtures must submit lists of ongoing health and safety studies being conducted by or for them, and unpublished studies known to them, but for which they do not have copies. In the case of studies not in their possession, the name and address of the person known to be in possession of the study must be provided.

Submission of lists of ongoing studies must be indexed by Chemical Abstract Service reporting number and accompanied by a cover letter which identifies the name, address, and job title of the submitting official and the name and address of the manufacturing or processing establishment on whose behalf the list is submitted. 40 C.F.R. § 716.35.

EPA Requests for Further Information

EPA also modified the confidentiality provisions of the Model Reporting Rule in its 1986 amendments. Prior to the amendments, persons who submitted health and safety studies as required by § 8(d) could claim all or part of the study as confidential. In response to what the Agency perceived as excessive claims of confidentiality and an apparent conflict with TSCA § 14, EPA amended the confidentiality provision. Now, only company name, address, financial statistics, and product codes can be claimed confidential along with the information permitted to be claimed confidential under TSCA § 14. 51 Fed. Reg. 32,724.

Following the initial reporting of copies and lists of studies, EPA may request the underlying data, preliminary reports of ongoing studies or submission of copies of studies listed by persons who actually possess studies, but which the manufacturer or processor who reported the existence of the studies did not possess. 40 C.F.R. § 716.40.

SECTION 8(e): SUBSTANTIAL RISK INFORMATION

Section 8(e) is unique among TSCA's § 8 reporting requirements. It is the only part of § 8 which is "self-actuating," and the only section which requires manufacturers to make their own subjective judgments as to the types of information required to be reported. Section 8(e) provides:

> Any person who manufactures, processes, or distributes in commerce a chemical substance or mixture and who obtains information which reasonably supports the conclusion that such substance or mixture presents a substantial risk of injury to health or the environment shall immediately inform the

Administrator of such information unless such person has actual knowledge that the Administrator has been adequately informed of such information.

Section 8(e) was promulgated by Congress in response to a concern that companies could legally withhold key information concerning the potential dangers of certain chemicals, such as kepone and vinyl chloride, from both the government and their workers. Section 8(e) was designed to ensure that information which indicates substantial risk receives prompt attention from EPA.

EPA has not issued regulations implementing § 8(e). The Agency has, however, issued guidance on § 8(e). Statement of Interpretation and Enforcement Policy, 43 Fed. Reg. 11,110 (1978) (1978 Policy Statement). EPA also provides some limited policy guidance through § 8(e) status reports, and from time to time through its monthly publication, "TSCA Chemicals in Progress." The status reports are a summary of EPA's initial review of submitted § 8(e) and "for your information" (FYI) reports and are available for public viewing in the OTS Public Reading Room at EPA Headquarters in Washington, D.C.[10] As of May 1989, EPA had received over 800 § 8(e) notices.

Who Must Report

Persons who manufacture, process, or distribute in commerce a chemical substance or mixture are subject to the § 8(e) reporting requirements. There is no small business exemption under § 8(e). A person who comes into receipt of information about a particular chemical which appears to constitute substantial risk information, but who does not manufacture, process, or distribute that chemical, is under no obligation to report the information. Thus, trade associations and testing laboratories generally are not subject to § 8(e). This information, however, is often reported to the Agency on a "for your information" (FYI) basis. EPA also will accept information on an FYI basis when the information may indicate some risk, but does not rise to the level of a "substantial" risk.

In the case of business entities, the chief executive officer, president, and any other officers responsible and having authority for the company's § 8(e) program must ensure that any substantial risk information is reported to the Agency. EPA considers a business entity to have obtained the information when any officer or employee "capable of appreciating the significance of that information" obtains it. 1978 Policy Statement at 11,111. EPA takes the position that such officers and employees are also individually subject to the § 8(e) reporting requirements and can incur civil and criminal liability for failure to report substantial risk information. A company can relieve its officers or employees of their obligation as individuals to report the information directly to the Agency by "establishing, internally publicizing, and affirmatively implementing procedures for employee submission and corporate processing of pertinent information." Id.

What Constitutes Substantial Risk Information

TSCA § 8(e) requires the reporting of information which "reasonably supports" the conclusion that the chemical substance or mixture presents a "substantial risk." Neither TSCA nor its legislative history, however, define the term "substantial risk" and EPA's 1978 Policy Statement and the Agency's Status Reports provide only limited guidance. According to EPA, a substantial risk is a risk of "considerable concern because of (a) the seriousness of

---

[10] Copies of individual EPA § 8(e) status reports are available from the EPA TSCA Assistance Office. Biennial "Blue Book" volumes of status reports are available from the TSCA Assistance Office in Washington, D.C., or from the National Technical Information Service in Springfield, Va., (703) 487-4600.

the effect . . . and (b) the fact or probability of its occurrence." 1978 Policy Statement at 11,111. The 1978 Policy Statement lists three categories of effects that are potentially serious enough to trigger reporting obligations: human health effects; environmental effects; and emergency incidents of environmental contamination.

Human Health Effects

EPA requires substantial risk information to be reported on the following human health effects: "(1) Any instance of cancer, birth defects, mutagenicity, death, or serious or prolonged incapacitation, including the loss of or inability to use a normal bodily function with a consequent relatively serious impairment of normal activities, if one (or a few) chemical(s) is strongly implicated. (2) Any pattern of effects or evidence which reasonably supports the conclusion that the chemical substance or mixture can produce cancer, mutation, birth defects or toxic effects resulting in death, or serious or prolonged incapacitation." Id. at 11,112. Any such effects definitely must be reported if they occur in humans and a particular substance or mixture is implicated. In addition, lesser effects on humans may be reportable if they may be preliminary manifestations of the more serious effects. Where the information indicates only that the chemical produces serious effects in laboratory animals, however, a potential reporter must determine whether the probability of exposure to humans or other organisms is sufficient to support the conclusion that the compound may present a substantial risk.

Exposure Factors

It is a generally recognized principle of risk assessment that risk is a function of both toxicity and exposure. As EPA has stated: "Both hazard and exposure must occur for there to be any risk. If there is virtually no exposure, EPA will be unable to conclude that the chemical may present an unreasonable risk." G. Timm, Remarks at The Toxicology Forum 1984 Annual Summer Meeting (July 16-20, 1984). EPA essentially disregards the requirement of exposure potential, however, when determining whether certain toxicity information reasonably supports a conclusion that a chemical may present a substantial risk. According to the Agency, the serious health effects listed in the 1978 Policy Statement "are so serious that relatively little weight is given to exposure; the mere fact the implicated chemical is in commerce constitutes sufficient evidence of exposure." 1978 Policy Statement at 11,111. In practice, EPA staff often interprets the standard of "relatively little weight" to mean "no weight."

The exposure issue is critical to the determination of reportability of toxicity information on chemicals that are at the research and development phase. Companies often perform or commission toxicity tests on chemicals before beginning commercial production.[11] If the results of these tests indicate that the chemical can cause a serious health effect, the company must determine whether there is sufficient potential for exposure to reasonably support the conclusion that the chemical presents a substantial risk.

In an oral question and answer session on § 8(e) that EPA sponsored on March 16, 1978, EPA officials apparently stated that the Agency did not want to receive information on R&D chemicals that never left the laboratory. EPA's current position, however, is that any test results that show that a chemical can produce a serious health effect will trigger § 8(e)

---

[11] There is no question that § 8 applies to chemicals in the R&D phase. Although § 8(f) limits the application of § 8 to chemicals that are manufactured, processed or distributed for a commercial purpose, EPA has asserted that manufacture, processing or distribution for R&D is done for a commercial purpose, and this view has been affirmed by the United States Court of Appeals for the Third Circuit. Dow Chemical Co. v. EPA, 605 F.2d 673 (1979).

reporting obligations even if the chemical never leaves the laboratory. EPA took that position in a January 1986 status report which arose from an FYI submission of preliminary results of a two-year oncogenicity study on a candidate pesticide that indicated that the chemical could cause liver tumors in mice. Status Report 8EHQ-0685-0583 S/8EHQ-1085-0583 S FLWP (EPA Jan. 7, 1986).

The FYI data submitter asserted that the test results did not trigger § 8(e) reporting obligations for several reasons, including the fact that the chemical was only produced in small quantities for laboratory research. EPA disagreed, and stated that the information should have been submitted as a formal § 8(e) report. The status report does not explain, however, how substantial risk can be inferred solely from toxicity data. EPA did not take any enforcement action against the company for failure to submit the information as a formal § 8(e) report.

Acute Toxicity Data

Although EPA has explicitly stated that instances of cancer, birth defects, mutagencity or death constitute reportable information under § 8(e), the Agency has also defined substantial risk in terms of serious or prolonged incapacitation and patterns of effects which indicate a potential for such. Serious or prolonged incapacitation, of course, can result from a chemical's acute toxicity. Thus, a frequent issue under § 8(e) is whether receipt of acute toxicity information triggers § 8(e) reporting obligations.[12/]

Acute studies are used to develop safety measures for those exposed to the chemical. These types of studies often can provide information on the mechanisms of toxicity and the structure/activity effect relationship for a particular class of chemicals. One of the most commonly performed acute toxicity tests is the LD_{50} (defined as the dose of the compound which causes 50 percent mortality in a given population). In responding to public comments concerning the reportability of routine tests such as the LD_{50}, EPA has stated that unknown effects occurring during a standard range-finding acute toxicity test may have to be reported if they meet the definition of substantial risk, but the Agency implies that results of routine tests may not be reportable if they do not show any unexpected effect. 1978 Policy Statement at 11,114 (response to Comment 14).

EPA also has addressed the question in its § 8(e) status reports. In § 8(e) Status Report 8EHQ-0984-0531-S, a company had provided summarized results from several acute in vivo toxicity studies of the fatty acid imidazoline, including skin irritation studies and an oral LD_{50}. In evaluating the reportability of this information under TSCA § 8(e), the Agency stated that "subject companies should consider such factors as the lethal dose, the pH of the tested chemical or mixture, the route of administration, the occurrence of unexpected effects (which could be determined via 'cage-side' observation or during necropsy), and the extent and pattern of the actual or potential exposure to the tested chemical or mixture. In general, when evaluating such information <u>for TSCA Section 8(e) reporting, the greater the acute toxicity, the less heavily a subject company should weigh the exposure to the tested chemical or mixture, and vice versa</u>." EPA, Office of Pesticides and Toxic Substances, Preliminary Evaluations of Initial TSCA Section 8(e) Substantial Risk Notices Jan. 1, 1983 - Dec. 31, 1984 at 226 (Mar. 1985) (emphasis added). In other status reports, EPA questioned whether submitted studies that showed moderate acute toxicity warranted formal § 8(e) submission. See status reports 8EHQ-0788-0742 (July 21, 1988); 8EHQ-0487-0665 S (May 14, 1987); 8EHQ-0487-0669 (May 7, 1987).

[12/] Acute toxicity has been defined as "the adverse effects occurring within a short time of [oral] administration of a single dose of a substance or multiple doses given within 24 hours." A.W. Hayes, Principles and Methods of Toxicology at 1-2 (1986).

EPA also has questioned whether all acute effects on humans must be reported. For example, a company submitted information concerning an employee's reaction to fumes generated during the heating and molding of a certain compound. The Agency concluded that this reaction need not have been reported under § 8(e). In Status Report EHQ-1283-0502, EPA stated that such information, together with "another triggering piece of information," could constitute reportable information under § 8(e), but "[c]onsidering that the acute human health effects reported in this submission do not appear to have been serious or to have resulted in prolonged incapacitation or serious impairment of normal bodily functions/ activities, EPA does not believe that the information warranted submission under Section 8(e) of TSCA." Status Report at 4.

It should be noted that, although EPA does not consider minor symptoms of acute human toxicity to be reportable under § 8(e), such information, if reported by an employee, consumer, or other person, constitutes an "allegation of a significant adverse reaction" under § 8(c) of TSCA.

Environmental Effects

Part V(b) of the 1978 Policy Statement contains the following requirements for reporting environmental effects data:

(b) <u>Environmental effects</u> -- (1) Widespread and previously unsuspected distribution in environmental media, as indicated in studies (excluding materials contained within appropriate disposal facilities).

(2) Pronounced bioaccumulation. Measurements and indicators of pronounced bioaccumulation heretofore unknown to the Administrator (including bioaccumulation in fish beyond 5,000 times water concentration in a 30-day exposure or having an n-octanol/water partition coefficient greater than 25,000) should be reported when coupled with potential for widespread exposure and any non-trivial adverse effect.

(3) Any non-trivial adverse effect, heretofore unknown to the Administrator, associated with a chemical known to have bioaccumulated to a pronounced degree or to be widespread in environmental media.

(4) Ecologically significant changes in species' interrelationships; that is, changes in population behavior, growth, survival, etc. that in turn affect other species' behavior, growth, or survival.

> Examples include: (i) Excessive stimulation of primary producers (algae, macrophytes) in aquatic ecosystems, e.g., resulting in nutrient enrichment, or eutrophication, of aquatic ecosystems.
>
> (ii) Interference with critical biogeochemical cycles, such as the nitrogen cycle.

(5) Facile transformation or degradation to a chemical having an unacceptable risk as defined above.

43 Fed. Reg. 11,112 (1978). EPA considers incidents of groundwater contamination to be included in the term "environmental media" and thus subject to § 8(e). EPA, TSCA Chemicals in Progress Bulletin (Mar. 1985) (available from the TSCA Assistance Office).

In contrast to human health effects, exposure plays a key role in determining the reportability of environmental effects. As EPA states in Part V of its 1978 Policy Statement: "[T]he remaining effects listed in Subparts (b) and (c) [environmental effects] below <u>must involve, or be accompanied by the potential for, significant levels of exposure</u> (because of general production levels, persistence, typical uses, common means of disposal, or other pertinent factors.)" 43 Fed. Reg. 11,111 (1978) (emphasis added).

Emergency Incidents of Environmental Contamination

Emergency incidents of environmental contamination also may be reportable under § 8(e) if they meet the criteria set forth in Part V(c) of EPA's 1978 Policy Statement: "<u>Emergency incidents of environmental contamination</u> -- Any environmental contamination by a chemical substance or mixture to which any of the above adverse effects has been ascribed and which because of the pattern, extent, and amount of contamination (1) seriously threatens humans with cancer, birth defects, mutation, death, or serious or prolonged incapacitation, or (2) seriously threatens non-human organisms with large-scale or ecologically significant population destruction." Id. at 11,112.

For example, incidents such as the accidental rupture of railroad tank cars containing hazardous chemicals, or the discovery of transformers leaking PCBs, are considered emergency incidents of environmental contamination. In most cases, § 103(a) of the Comprehensive Environmental Response, Compensation and Liability Act (CERCLA) requires responsible parties to report such incidents immediately to the National Response Center for Oil and Hazardous Material Spills, and under the CERCLA implementing regulations, a written follow-up report may be required. 42 U.S.C. § 9603 (1982 & Supp. V 1987); 40 C.F.R. pt. 302. These reports may obviate the need for a separate submission under § 8(e), since under § 8(e) no report is required if the responsible party "has actual knowledge that the Administrator has been adequately informed of such information."

Nature and Sources of Information Which Reasonably Supports the "Conclusion" of Substantial Risk

EPA considers that information from designed, controlled studies as well as reports concerning, and studies of, undesigned, uncontrolled circumstances can "reasonably support" a finding of substantial risk. In fact, EPA has indicated that even informal communications between companies can trigger § 8(e) reporting obligations, if a company thereby "obtains" substantial risk information on a chemical it manufactures, processes or distributes in commerce. In § 8(e) Status Report 8EHQ-0585-0572/8EHQ-0985-0572 Followup (Status Report Followup), EPA stated that a company incurred reporting obligations when its official heard by telephone that another company had found teratogenic effects while doing a study of the first company's chemical.

Designed, Controlled Studies

Studies which reasonably support a substantial risk finding include: (1) <u>in vivo</u> experiments and tests, (2) <u>in vitro</u> experiments and tests, (3) epidemiological studies, and (4) environmental monitoring studies. Status Report Followup. According to EPA, conclusions concerning effects on humans can be obtained directly or inferred from these designed studies. In certain instances, even preliminary results from incomplete studies would trigger § 8(e) reporting obligations, and the responsible parties should not await final results.

Corroborative information is relevant in determining whether a positive <u>in vitro</u> study constitutes reportable information under § 8(e). The existence of corroborative information is particularly important in evaluating the results of short term <u>in vitro</u> tests, such as the Ames, Sister Chromatid Exchange, and mouse lymphoma assays, all of which give a high percentage of false positive results. The Agency, however, has failed to provide any guidance on what constitutes corroborative information. Many submitters have reported uncorroborated positive short term test results on an FYI basis, asserting that these results provided "limited" evidence of risk but would not reasonably support the conclusion that the chemical poses a substantial risk. EPA staff has suggested in informal guidance that prospective data submitters should consider submitting positive short-term results of a single in vitro test as a § 8(e) submission if the test results are unequivocal and there is substantial potential for exposure to the chemical.

Undesigned, Uncontrolled Circumstances

Reports concerning and studies of undesigned or uncontrolled circumstances may also give rise to § 8(e) reporting obligations. Such reports and studies can include: (1) medical and health surveys, (2) clinical studies, and (3) reports concerning and evidence of effects in consumers, workers, or the environment. Id. For example, a company medical survey may reveal a pattern of health effects among employees that would give rise to a § 8(e) reporting obligation even where none of the individual reports would be reportable.

EPA anticipates that most reports triggered by studies of undesigned, uncontrolled circumstances will be submitted because the study reveals patterns of health effects and their correlation with exposure to a chemical. In some circumstances, however, such studies may reveal a more serious effect and implicate a chemical, triggering reporting obligations even in the absence of any pattern of effects. EPA has stated that "a single instance of cancer, birth defects, mutation, death, or serious incapacitation in a human would be reportable if one (or a few) chemical(s) was strongly implicated." Id.

Exemption for Corroborative Information

EPA's 1978 Policy Statement provides that substantial risk information need not be reported if it "[i]s corroborative of well-established adverse effects already documented in the scientific literature and referenced [by certain abstracting services]." 1978 Policy Statement at 11,112.

EPA's 1978 Policy Statement, however, does not provide any further definition of the phrase "corroborative of well-established adverse effects." Indeed, the public record prior to the publication of the 1978 Policy Statement is silent on the definition of the phrase. Not until 1984 did EPA announce its interpretation of the phrase. In a § 8(e) enforcement proceeding, EPA alleged that a company violated § 8(e) by failing to submit the results of a skin painting study on rodents which indicated that a compound was carcinogenic when applied to rodent skin. The company did not report the study because company officials believed that the study results merely corroborated the previously known carcinogenicity of the compound. EPA disagreed, and announced a new standard for determining when information is corroborative: "[I]n order for new health effects information to be corroborative, the information must verify or confirm previously reported and/or published information that the subject chemical substance or mixture when administered at a specific dose by a specific route produced a specific toxic effect (e.g., squamous cell carcinoma) in a specific target organ (e.g., skin) in a specific species or strain of species." EPA, Office of Pesticides and Toxic Substances, Preliminary Evaluations of Initial TSCA Section 8(e) Substantial Risk Notices Jan. 1, 1983 - Dec. 31, 1984 at 154 (Mar. 1985).

Under this five-part test, it appears that new information will be considered corroborative only if it essentially replicates a prior finding. Under this test, a new study would "rarely, if ever, corroborate" a prior chronic study because some parameter would almost always be different. The EPA enforcement action was settled prior to a scheduled hearing. Under the terms of the settlement, the company continues to maintain that the study in question was merely corroborative of well-known adverse effects, and no court has had an opportunity to rule on EPA's newly asserted standard.

Submission of § 8(e) Information by Trade Associations

Many trade associations sponsor and report studies on chemicals although they have no obligation to report under § 8(e) because they are not manufacturers, processors, or distributors. The information is generally communicated to their member companies who are subject to § 8(e). EPA has addressed the submission of § 8(e) information by trade associations in a § 8(e) status report:

> [T]he obligation to immediately report substantial risk information would apply to an organization/association such as SPI, CMA, API, CIIT, etc. only if such organization/association engages in the manufacture, import, processing, or distribution of the chemical substance or mixture about which substantial risk information has been obtained. However, it is more likely that a Section 8(e) reporting obligation would be incurred by a member (or non-member) company that obtains the results (including preliminary results) from studies conducted or sponsored by an organization/association, if the obtained results meet the reporting criteria outlined in the TSCA Section 8(e) policy statement and pertain to a chemical substance or mixture that the company manufactures, imports, processes, or distributes in commerce.

EPA, Status Report 8EHQ-0185-0543 at 3 (Mar. 6, 1985).

FYI Submissions Versus § 8(e) Submissions

Many companies or persons have sent toxicological and other information to EPA on an FYI basis. These FYI submitters have determined that their information, while indicating some degree of risk, does not indicate a "substantial" risk. For example, acute toxicity and mutagenicity test results often are submitted on an FYI basis. These FYI submissions are available for review in the OTS Public Reading Room at EPA Headquarters.

The FYI submissions are processed by EPA in a manner which is almost identical to a § 8(e) notice, except that a status report is not prepared on an FYI submission. Upon receipt of an FYI notice, the OTS § 8(e) staff reviews it to determine whether the submission should have been submitted under § 8(e). If the staff determines that the information should have been submitted pursuant to § 8(e), the submitter is notified by EPA and given twenty working days in which to respond to EPA's findings and to explain why the information was not submitted pursuant to § 8(e). EPA then evaluates the response and, if the Agency still determines that the information constitutes § 8(e) information, the FYI is "converted" into a § 8(e) notice and enters the § 8(e) notice processing system. The history of the submission, however, also is forwarded to the Office of Compliance Monitoring for evaluation of the situation and possible enforcement action. To date, EPA has not prosecuted anyone for submitting information as an FYI that EPA determined should have been submitted under § 8(e), if the submission was made within the § 8(e) reporting deadline, i.e., within fifteen working days after the person received the information.

When to Submit § 8(e) Reports

TSCA § 8(e) requires that substantial risk information be submitted "immediately" to the Agency. Neither TSCA nor its legislative history define the term. In its 1978 Policy Statement, EPA stated that most § 8(e) information must be submitted "not later than the 15th working day after the date the person obtained such information." 43 Fed. Reg. 11,111 (1978). Reporting is not to be delayed while "conclusive" information is developed. EPA has recognized that the evidence generally will not be conclusive: "Such evidence will generally not be conclusive as to the substantiality of the risk; it should, however, <u>reliably ascribe the effect to the chemical.</u>" Id. at 11,112 (emphasis added).

Emergency incidents, however, are to be reported initially by telephone to the appropriate EPA region within twenty-four hours. (See Figure 1, EPA Regional 24-Hour Response Centers.) A written submission must then be filed with the OTS Document Control Officer at EPA Headquarters in Washington, D.C. within fifteen working days. 1978 Policy Statement at 11,113.

Confidentiality Claims and § 8(e) Reports

Persons submitting a § 8(e) notice may claim some or all of the information as confidential. Any information claimed as confidential will be disclosed only to the extent authorized in 40 C.F.R. pt. 2. If no information is claimed confidential, the entire notice will be placed in the public file.

In order to claim information confidential, the submitter must specifically indicate which information is claimed confidential by labeling each page "proprietary," "trade secret" or "confidential." If a submitter claims any information in the notice confidential, a second ("sanitized") copy of the notice must be submitted from which the confidential information has been deleted. This sanitized copy is then placed in the public file.

EPA Processing of § 8(e) Notices

EPA has developed an elaborate system for processing and reviewing § 8(e) notices. The review is coordinated by the Chemical Screening Branch of the Existing Chemicals Division. Within thirty days of receipt of a § 8(e) notice, EPA will review the notice, prepare a § 8(e) status report, and recommend any additional follow-up actions. This thirty-day time limit is self-imposed. TSCA imposes no statutory deadline for EPA's response to a § 8(e) notice.

A § 8(e) notice initially is received by the Document Control Officer of OTS, who logs the submission into the OTS § 8(e) tracking system, assigns a document control number, reviews the submission for confidential information, and contacts the § 8(e) Coordinator. The § 8(e) notice (or its sanitized version) is forwarded to the OTS public reading room and a copy of the original, confidential version is also routed to the § 8(e) Coordinator.

The § 8(e) Coordinator functions in the same way as a Program Manager in the New Chemicals Review Program; the Coordinator is responsible for integrating the review of the § 8(e) notice. A review of a § 8(e) notice includes an evaluation of the applicability of § 8(e) to the information contained in the notice, review by scientists of the health, exposure, or ecotoxicity issues posed by the information, and a determination of any additional actions which must be taken in response to the information. The § 8(e) Coordinator incorporates the reports of the various scientists into a § 8(e) status report. Essentially a § 8(e) status report is an internal memorandum from the § 8(e) Coordinator to the Branch Chief of the Chemical Screening Branch.

The completed status report is then transmitted to the submitting company. In this letter, EPA may request additional information concerning the circumstances surrounding the reported information. For example, EPA may request a description of any actions taken by the company in response to the § 8(e) information, such as modifications to its material safety data sheets. Approximately seven days later, the status report is forwarded to the OTS Public Reading room at EPA Headquarters for inclusion in the § 8(e) public file. This brief delay in forwarding the status report to the OTS Public Reading room is designed to give the submitting company time to receive the status report first. In addition to transmitting a copy to the public reading room, the § 8(e) Coordinator may send copies to the following EPA offices and federal agencies or departments: National Library of Medicine; Occupational Safety and Health Administration; National Institute for Occupational Safety and Health; Consumer Product Safety Commission; Food and Drug Administration; National Toxicology Program; Office of Research and Development/EPA; Office of Water/EPA; Office of Solid Waste and Emergency Response/EPA; Risk Management Branch/ECAD/OTS/EPA; Office of Air and Radiation/EPA; and Office of Pesticide Programs/EPA.

DEVELOPING A REPORTING AND RETENTION PROGRAM

In order to comply with the reporting and recordkeeping provisions of § 8, companies must develop an internal § 8 compliance program. While each compliance plan must be tailored to a company's individual structure and needs, there are five key elements which must exist in any sound § 8 program.

First, a § 8 compliance program must be an ongoing activity that is well integrated into regular management and operations activities. Second, it is extremely important for the program to have very visible support from the company's top management. This emphasizes the importance of § 8 compliance to overall corporate success. Many companies achieve this top level support by having the company's chief executive officer sign major communications regarding § 8 compliance. Third, the compliance system must allow a two-way flow of information between people who track EPA developments and people who follow company activities. Fourth, people who are responsible for knowing about EPA requirements and those who actually will implement the compliance program in plants, laboratories, or other sites should be involved in developing or revising the program. Finally, it is important to periodically remind people about their TSCA § 8 responsibilities through memos, talks, or meetings.

Specific Compliance Suggestions

Section 8(a)

Under TSCA § 8(a), EPA may request several kinds of information regarding a chemical, including the amount a company produces, the number of people exposed, or the chemical's health effects. Someone in the company must be responsible for checking the Federal Register for new EPA reporting requirements. Some companies delegate this responsibility to someone in their legal department, others to someone in their regulatory affairs department.

Once aware of a new EPA requirement, this person must identify and inform those persons in the company who likely have information about the chemical subject to the new EPA requirements. The way this is done will depend on the company's structure. In a smaller company it may be enough to identify and inform the plant manager who deals with the chemical. Larger companies sometimes have a steward assigned to each of their products, in which case that person should be informed. If a company maintains a computerized inventory of chemicals that includes information regarding where chemicals are being produced, the affected chemicals can be flagged. In other companies, it may be more convenient to identify people who have information about the chemical from the company's organizational chart. Regardless of how they are identified, those people who are likely to have the information needed in a § 8(a) report should be informed and required to gather the data. Someone also must be designated to review the information gathered and prepare it for submission to EPA.

Section 8(b)

Companies often combine Inventory reporting responsibilities with § 8(a) responsibilities. The important thing to remember is that in requiring an Inventory update, EPA may ask for information on plant production for a past period (e.g., the last year). Because of this, it is worthwhile to incorporate plant level recordkeeping into a company's § 8(a) and § 8(b) compliance program.

Sections 8(c) and 8(e)

Since companies can use similar systems for complying with both §§ 8(c) and 8(e), they are discussed together here. First, the company must have a system in place for gathering § 8(c) and § 8(e) information. This may be accomplished by having a system of product stewardship or responsibility built into the company's structure. Part of the product steward's responsibility would be to gather § 8(c) allegations and information regarding the product's health or environmental risks. Another way § 8(c) and § 8(e) information can be gathered is to have at each location in the company a TSCA coordinator who is trained in TSCA compliance and in turn is responsible for training people at their locations in how to identify either § 8(c) allegations or § 8(e) substantial risk information. This TSCA coordinator also would have the responsibility of reporting to the corporate committee or committees responsible for § 8(c) and § 8(e).

Second, one or two committees should be formed at the corporate level to evaluate § 8(c) and § 8(e) information. At a minimum, the company's medical, industrial hygiene, toxicology, operations, and legal departments should be represented on such committees. These committees should have two major responsibilities: (1) educating plant level product stewards or TSCA coordinators regarding the requirements of § 8(c) and § 8(e) and the company's compliance policy, and (2) reviewing submitted information to determine whether it is a record that should be retained under § 8(c) or information that must be reported under § 8(e). The committee should see that several educational activities are incorporated into company practice. Training in § 8 compliance should be included in the initial training of product stewards or TSCA coordinators. The committee could provide continuing training at yearly meetings. The § 8(c) and § 8(e) committees should send annual out memos, signed by the Company's CEO, reminding people who may get § 8(c) or § 8(e) information of their reporting responsibilities and explaining both the company's compliance program and EPA's reporting requirements. The committee responsible for § 8(c) should meet periodically (monthly, quarterly, or semi-annually) to review § 8(c) submissions and determine which are § 8(c) allegations that should be retained in the company's files. Because § 8(e) information must be reported within fifteen working days, the § 8(e) committee must meet on an *ad hoc* basis as potentially reportable information is discovered. The § 8(e) committee also should periodically review § 8(c) allegations to determine whether there is any pattern of effects indicating that a chemical may pose substantial risk of harm to health or the environment reportable under § 8(e).

Section 8(e) also requires prompt reporting of emergency incidents. Because a committee generally could not review and respond to such an incident quickly enough, the company should designate an individual as responsible for deciding whether an incident should be reported as an emergency incident. This person may be a committee member or someone at the plant level. This person also should know the company's emergency reporting responsibilities under other environmental laws. These laws may require reporting to the National Response Center or to state and local emergency response committees.

Section 8(d)

In order to ensure compliance with § 8(d), someone in the company must monitor the Federal Register for new health and safety study reporting requirements. This could be the same person who reviews the Federal Register for TSCA § 8(a) and § 8(b) requirements. People who are responsible for the chemical must then be informed of the need to report. Again, these people could be identified via a computerized chemical inventory that identifies where the chemical is manufactured, or through a diagram of corporate responsibilities. A company also might consider maintaining a data base of all toxicological studies and a data base of all industrial hygiene studies. The data bases could be labeled so they can retrieve studies by the chemical substances or mixtures examined. If this system is followed, the data bases should be searched and the responsible people should be informed of the new reporting obligation.

Once the studies are gathered, they should be reviewed by legal or regulatory affairs personnel to see if they are reportable under TSCA and by products personnel to see if they contain confidential business information. Someone, of course, must be responsible for submitting the reports to EPA.

EPA also requires ongoing reporting under TSCA § 8(d). The company person responsible for following the Federal Register should remind company personnel each year about their ongoing reporting obligation.

FIGURE 1. EPA REGIONAL 24-HOUR RESPONSE CENTERS

Region I	(Maine, Rhode Island, Connecticut, Vermont, Massachusetts, New Hampshire) 617/223-7265
Region II	(New York, New Jersey, Puerto Rico, Virgin Islands) 201/548-8730
Region III	(Pennsylvania, West Virginia, Virginia, Maryland, Delaware, District of Columbia) 215/597-9898
Region IV	(Kentucky, Tennessee, North Carolina, South Carolina, Georgia, Alabama, Mississippi, Florida) 404/881-4062
Region V	(Wisconsin, Illinois, Indiana, Michigan, Ohio, Minnesota) 312/353-2318
Region VI	(New Mexico, Texas, Oklahoma, Arkansas, Louisiana) day: 214/655-2210; 24-hour: 214/655-2222
Region VII	(Nebraska, Iowa, Missouri, Kansas) 913/236-3778
Region VIII	(Colorado, Utah, Wyoming, Montana, North Dakota, South Dakota) 303/844-1788
Region IX	(California, Nevada, Arizona, Hawaii, Guam) 415/974-8131
Region X	(Washington, Oregon, Idaho, Alaska) 206/442-1263

III. NEW CHEMICAL REVIEW: OVERVIEW AND EXEMPTIONS

Any person intending to manufacture or import a chemical substance first must determine whether it is listed on the TSCA Inventory. If it is listed, then manufacture or importation may commence immediately. If the chemical substance is not listed on the Inventory, then the manufacturer/importer must determine whether the chemical substance is excluded altogether from regulation under TSCA or whether it is exempt from the premanufacture notification (PMN) requirements under TSCA § 5(h). If the chemical substance is neither excluded nor exempted, the prospective manufacturer or importer must comply with the PMN requirements before commencing those activities.

This section discusses the PMN requirements and the exemptions from those requirements. Section IV provides detailed guidance on complying with the PMN requirements, and section V discusses reporting requirements for § 5(e) consent orders and Significant New Use Rules (SNURs).

OVERVIEW OF TSCA § 5

TSCA § 5 provides EPA the authority to evaluate and, in certain instances, to regulate new chemical substances. More specifically, § 5(a) requires a manufacturer (which by definition includes an importer) of a new chemical substance to file a PMN at least ninety days prior to commencement of commercial production (or importation). This PMN review period provides EPA its first (but certainly not its only) opportunity to evaluate new chemical substances. The PMN provision reflects the House and Senate conferrees' view that "the most desirable time to determine the health and environmental effects of a substance, and to take action to protect against any potential adverse effects, occurs before commercial production begins. Not only is human and environmental harm avoided or alleviated, but the cost of any regulatory action in terms of loss of jobs and capital investment is minimized." H.R. Conf. Rep. No. 1679, 94th Cong., 2d Sess. 65 (1976).

PMN REQUIREMENTS

The contents of a PMN are described very specifically in TSCA § 5(d). The PMN must contain: (1) certain information described in § 8(a)(2) such as the identity of the chemical, categories of use, amounts manufactured, byproducts, employees exposed and the manner or method of disposal to the extent known or "reasonably ascertainable"; (2) any test data related to the chemical's effects on health or the environment in the submitter's possession or control; and (3) a description of any other data concerning the health and environmental effects of the chemical, insofar as they are known to or "reasonably ascertainable" by the submitter. In using the phrase "reasonably ascertainable," Congress intended to create an "objective" standard which includes "information of which a reasonable person similarly situated might be expected to have knowledge."

The policy underlying TSCA is that manufacturers and processors of chemical substances should bear the responsibility for developing adequate data regarding their effects on health and the environment. Section 5(b)(2)(B) thus obligates the PMN submitter to produce such data as he believes will show that the manufacture, processing, distribution in commerce, use, and disposal of the new chemical substance or any combination of such activities "will not present an unreasonable risk of injury to health or the environment." Nevertheless, § 5 does not expressly authorize the Administrator to require or obligate the PMN submitter to produce specific tests with a PMN, except where a chemical substance is subject to a rule promulgated under § 4. Thus, unlike comparable laws outside the United States, TSCA does not require a bare set of premarket data on a new chemical.

EPA's regulations implementing TSCA § 5 were promulgated in a seven-year rulemaking ending in 1986.[13/] Under the regulations, EPA requires that the PMN be submitted on a standard form (EPA Form No. 7710-25) and that the PMN contain the name and address of the submitter, the chemical identity of the substance, the estimated production volume, the type of use, and information concerning exposure of workers, users, and consumers, and concerning release to the environment. 40 C.F.R. § 720.40. (See section IV.) After a PMN is submitted, the submitter may withdraw it voluntarily from review at anytime. Although the Agency must be notified in writing of the submitter's intention to withdraw the PMN, the submitter need not furnish any explanation for the withdrawal.

PMN Review Period

Under § 5(a), EPA must review the PMN within ninety days. EPA may extend the review period for an additional ninety days under § 5(c) if it has "good cause." Also, the submitter and the Agency may suspend the review period periodically or indefinitely by mutual consent. During its PMN review EPA must assess the potential risks associated with the manufacture, processing, distribution, use, and disposal of the new substance based upon information supplied by the PMN submitter and available from various Agency data bases and the scientific literature, and ultimately based upon the Agency's own professional judgment.

If EPA takes no regulatory action on the PMN within the 90-day (or, if extended, 180-day) review period, the submitter may commence commercial manufacture or importation forthwith and without the need for prior Agency approval. Within thirty days of commencing manufacture or importation, the manufacturer or importer must file a Notice of Commencement of Manufacture or Import (NOC). 40 C.F.R. § 720.102. The NOC certifies that commercial manufacture and/or importation actually has occurred. After receiving an NOC, EPA will add the PMN substance to the Inventory, and the new chemical will then become an "existing" chemical under TSCA.

Submitters' Post Review Duties

Notice of Commencement of Manufacture

If EPA fails to take any action on the new chemical substance, the submitter is free to manufacture or import the substance for commercial purposes on the ninety-first day after PMN submission. Within thirty days of first beginning this commercial manufacture or importation, the submitter must notify EPA of this fact. 40 C.F.R. § 720.102(b). He must submit this "notice of commencement of manufacture" (NOC) in writing and must provide the specific chemical identity, PMN number, and the date when commercial manufacture actually commenced.

This NOC constitutes a certification that commercial manufacture or importation has actually occurred; thus, it cannot be submitted in advance. In addition, if the submitter originally claimed the chemical identity as confidential and wants this confidentiality retained, he must substantiate this claim in accordance with the procedures set forth in 40

[13/] Regulations implementing these notice provisions were proposed on January 10, 1979 (44 Fed. Reg. 2242), reproposed on October 16, 1979 (44 Fed. Reg. 59,764), and promulgated in final form on May 13, 1983. Provisions of the rule were later clarified on September 13, 1983 (48 Fed. Reg. 41,132). The effective date of the final regulations was stayed to permit EPA to review several provisions of the rule in response to public comment. The final rule became effective on October 26, 1983. Revisions were proposed on December 27, 1984 (49 Fed. Reg. 50,201) and became final on April 22, 1986 (51 Fed. Reg. 15,096).

C.F.R. § 720.85(b) when filing his NOC. Unless the manufacturer provides this substantiation, EPA lists the specific chemical identity on the public TSCA Inventory.

Recordkeeping

The submitter must maintain records of the information contained in the PMN or exemption application for a period of five years from the date of commencement of manufacture. 40 C.F.R. § 720.78(a). In particular, the submitter must maintain documentation pertaining to the production volume for the first three years of manufacture and to "other data . . . in the submitter's possession or control." Id.

EXCLUSIONS FROM PMN REQUIREMENTS

The PMN requirements apply to a "new chemical substance" and, once an applicable rule is promulgated, to a "significant new use" of an existing chemical substance. The statutory definition of "chemical substance" excludes any mixture, any pesticide as defined by the Federal Insecticide, Fungicide and Rodenticide Act, and any food, food additive, drug, cosmetic, device as defined by the Federal Food, Drug and Cosmetic Act, various nuclear materials regulated under the Atomic Energy Act, and any tobacco or tobacco product. See TSCA § 3(2)(B). Thus, by definition, these substances are excluded from TSCA jurisdiction and, as such, are not subject to the PMN requirements. The PMN requirements nevertheless may apply to such "excluded" substances, if they also are intended for a "TSCA use."

EXEMPTIONS FROM PMN REQUIREMENTS

TSCA § 5(h) establishes exemptions from the PMN requirements for chemical substances intended for test marketing, research and development (R&D), and other uses. These exemptions apply to chemicals subject to TSCA jurisdiction (i.e., those not subject to specific exclusion). Some exemptions like the test marketing exemption require prior application and EPA approval. Others, like R&D, involve only the company's judgment that the exemption applies.

To obtain a test marketing exemption (TME) under § 5(h)(1), the manufacturer must file an application with EPA demonstrating that such marketing will not present an unreasonable risk of injury to health or the environment. The Agency expects that a TME application will include much the same information as a PMN. Thus, the TME is a curious hybrid. As with the PMN, the Administrator must act quickly on the TME -- he must approve or deny the application within forty-five days of its receipt. Unlike the PMN, the Agency must "approve" the TME before test marketing commences. Since the applicant for a TME must meet the requirements of a PMN, the benefits of a TME are limited to the satisfaction which comes from receiving an Agency "approval" and entering the market forty-five days earlier.

By contrast with the TME, the R&D exemption under § 5(h)(3) requires no application. To qualify, a manufacturer need only comply with certain criteria: (1) the use must be solely for R&D purposes; (2) the manufacture must be limited to "small quantities" no greater than reasonably necessary for such purposes; and (3) handling must take place under the supervision of a technically qualified individual. ("Small quantities" is not defined by statute or by regulation.) Hence, low volume alone does not establish entitlement to the R&D exemption. Moreover, the manufacturer or importer must notify all persons engaged in such experimentation, research, or analysis about any risk to health which he is reasonably certain may be associated with such chemical.

Like the TME, the "zero exposure" exemption under TSCA § 5(h)(5) requires an application to be filed with EPA. The exemption applies to a new chemical substance "which exists temporarily as a result of a chemical reaction" and to which "there is no, and will not be, human or environmental exposure." To date, EPA has not issued any regulations implementing TSCA § 5(h)(5) and has not granted any zero exposure exemptions.

Test Market Exemption

TSCA § 5(h)(1) authorizes the Administrator to exempt a new chemical substance from the PMN requirements when it is manufactured for test marketing purposes if the Administrator determines that the proposed test marketing activity will not present an unreasonable risk to human health or the environment. The TME permits a company to assess the commercial viability of a new chemical and to receive customer feedback on product performance before filing a PMN. The test marketer must apply for this exemption and must demonstrate that the proposed activity is legitimate test marketing which will not present an unreasonable risk. EPA must publish in the Federal Register both its receipt of the TME application and its decision to grant or deny the application.

EPA reviews a TME application in essentially the same manner as it does a PMN. (See section IV.) Yet under TSCA § 5(h)(6), EPA must review a TME application within forty-five days of its receipt. According to EPA, however, the Agency's failure to complete its review of a TME application within forty-five days does not constitute an automatic approval. Rather, unlike a PMN submitter, a TME applicant must await EPA approval prior to initiating activity.

EPA regulations implementing the TME are published at 40 C.F.R. § 720.38. They provide that a TME application need not be submitted on a PMN form; rather, it may be submitted in the form of a letter. The application, however, must contain the following information: all existing data regarding health and environmental effects, including physical/chemical properties or, in the absence of toxicity data, a discussion of structural analogues; the amount to be imported/manufactured; the number of persons who may be exposed; estimated routes and durations of exposures; and a description of test marketing activity, with particular attention to how the proposed activity differs from R&D and full scale commercial production.

Research and Development Exemption

TSCA § 5(h)(3) exempts from PMN and SNUR requirements small quantities of new chemicals used solely for R&D under the supervision of a technically qualified individual, if the manufacturer or importer notifies persons engaged in R&D of any health risks associated with the substance. Unlike the other exemptions under § 5(h), the manufacturer or importer need not apply for the R&D exemption.

EPA regulations establishing requirements for the R&D exemption appear at 40 C.F.R. § 720.36. EPA requires that manufacturers or importers supervise the R&D with a technically qualified individual, evaluate any potential risks associated with the R&D substance, notify persons involved in the R&D of those risks, and maintain certain records of their R&D activity. In evaluating risks, the manufacturer and importer must consider all health and environmental effects data in their "possession or control." This includes information in the files of agents and employees engaged in the R&D and marketing of the new chemical. When R&D activity is conducted in laboratories using prudent laboratory conditions, this risk assessment need not be performed.

Manufacturers and importers must maintain records pertaining to their risk evaluations of each new R&D chemical, the nature and method of their notification of any potential risk and, if they distribute the R&D chemical, the identity and amount of the distributed chemical and the identify of the recipient. In addition, those who manufacture or import an R&D substance in quantities greater than 100 kilograms per year must maintain records of the chemical identity, production volume, and disposition of the substance.

Evaluation, Notification, and Recordkeeping of Risks

Manufacturers and processors must evaluate information in their possession and control to determine whether R&D substances present any potential risk to human health or the environment, and must notify those engaged in R&D of any such risks. TSCA § 5(h)(3); 40 C.F.R. §§ 720.36(b)(1)(i), .36(c)(1). Laboratories using "prudent laboratory practices" are exempt from the risk evaluation and notification requirements. Similarly, EPA does not require risk evaluation and notification for the distribution of R&D substances between two laboratories using prudent laboratory practices.

The Agency has not published regulations or guidelines specifying what constitutes "prudent laboratory practices." EPA recommends, however, that manufacturers and processors follow standard procedures to control exposure and consult the following handbooks for guidance: National Research Council, Prudent Practices for Handling Hazardous Chemicals in Laboratories (1981); and U.S. Department of Health and Human Services, NIH Guidelines for the Laboratory Use of Chemical Carcinogens (May 1981). Exemptions Bulletin at 8.

EPA summarized the risk evaluation and notification requirements in the following table:

REQUIREMENTS FOR NOTIFICATION OF RISKS

Who must be Notified	Method of Notification	Content of Notification	Authority
Persons employed by a Firm for R&D work	Any appropriate means	Risks associated with R&D substance*/	720.36(a)(2)
Persons not in a firm's employ to whom it distributes R&D substances	In writing	Risks associated with R&D substance*/; Requirement that substance be used only for R&D	720.36(c)(2)

*/ In the case of workers in laboratories using prudent laboratory practices, or distribution from one such laboratory to another, EPA does not require evaluation of risks.

Exemptions Bulletin at 8.

-193-

The R&D rule requires that records be kept documenting compliance with either the risk evaluation and notification provisions or prudent laboratory practices. EPA summarized these requirements in the following table:

RECORDKEEPING REQUIREMENTS

Who Must Keep Records	Requirement	Authority
All manufacturers using the R&D exemption (except labs using prudent lab practices)	Copies of or citations to information reviewed to determine risks; Documentation of the nature & method of risk notification	720.78(b)(1)(i) & (ii)
Labs using prudent lab practices	Documentation of prudent lab practices	720.78(b)(1)(iii)
Manufacturers distributing R&D substances to persons outside their employ	Names and addresses of those who received substance; Identity of substance; Amount distributed; Copies of written notification	720.78(b)(1)(iv)
Producers of more than 100 kg/year	Record of identity of substance; Production volume Disposition	720.78(b)(2)

Exemptions Bulletin at 9.

R&D Pesticides

As a matter of policy, EPA has expressly exempted new chemicals undergoing R&D for possible use as pesticides from the risk assessment, notification, and recordkeeping requirements of the new R&D rule. 40 C.F.R. § 720.36(g). In order to qualify for this exemption, "the exclusive intention of the subsequent R&D activities of the person manufacturing the new substance and conducting the R&D activities [must be] to develop the substance as a pesticide." 51 Fed. Reg. 15,098 (1986). A new pesticide chemical may qualify for this exemption, however, "even if the potential properties of the substances as a pesticide are unknown at the time it is first manufactured." Id.

A manufacturer can show an exclusive pesticidal intent by having a patent for its use as a pesticide, or simply by operating as a company or laboratory devoted entirely to pesticide R&D. By contrast, a company with a broad range of marketing activities would need other positive evidence to establish the prerequisite exclusive pesticide intent. Where the R&D activity is not solely for pesticide purposes, the manufacturer or importer must fulfill the risk assessment, notification, and recordkeeping requirements of the R&D rule -- until the submission of an experimental use permit, a registration application, "or other activities . . . which provide evidence of exclusive pesticide intent." Id.

Although exempted from the requirements of the R&D rule, EPA considers that pesticide R&D remains subject to other applicable TSCA provisions, especially TSCA § 8(e). The § 8(e) provision requires notification to EPA of information concerning substantial risks associated with the candidate pesticide. (See section II.) Furthermore, although the preamble to the R&D rule implies that R&D pesticide intermediates and R&D pesticide inert ingredients may qualify for the R&D exemption for pesticides, the Agency apparently did not intend this result. Rather, EPA permits only the pesticide active ingredient to qualify for this exemption from the requirements of the R&D rule; the R&D inerts and R&D intermediates remain subject to the risk assessment, notification, and recordkeeping requirements.

Low Volume Exemption

Under TSCA § 5(h)(4), EPA has exempted certain low volume chemicals from the full PMN requirements by providing an expedited twenty-one day review. 40 C.F.R. § 723.50. A manufacturer or importer who intends to produce or import a new chemical substance in quantities of 1,000 kilograms or less per year may apply for the low volume exemption (LVE). This LVE is available only to one manufacturer for each new chemical. Therefore, once EPA has granted an LVE for a chemical, no other manufacturer can qualify for an LVE for that same chemical. A second manufacturer must submit a PMN.

LVE applicants need not use the standard PMN form; they need only identify that their submission is an LVE application. The LVE application must include the following: (1) the chemical identity (including its impurities); (2) a description of its use; (3) the site of its manufacture; (4) any test data on the chemical in the submitter's possession and control, and (5) a certification that it will be manufactured or imported in compliance with the requirements of the LVE.

The information required for an LVE differs from that for a full PMN. Unlike the full PMN, the LVE need not contain information on worker and user exposure or on exposure controls. In the absence of such information, however, EPA will make "reasonable assumptions" about exposure and such assumptions may not reflect actual exposure conditions. 50 Fed. Reg. 16,480 (1985). Hence, for all but toxicologically innocuous new chemicals, the LVE applicant may wish to supply such exposure information anyway in order to ensure the prerequisite "no unreasonable risk" finding by EPA.

EPA Review

In order to grant an LVE, EPA must determine that the chemical substance will "not present an unreasonable risk of injury to health or the environment." TSCA § 5(h)(4). EPA makes this determination based on the information contained in the LVE application. If the Agency concludes that the substance itself or a "reasonably anticipated metabolite or environmental transformation product of it" may cause either serious acute or chronic health effects or significant environmental effects, EPA will deny the LVE application. 40 C.F.R. § 723.50(d)(1).

The site of manufacture and specific use of the substance are critical to the Agency's risk determination. Since a change in either the manufacturing site or the use could cause substantially different exposures and enhanced risks, EPA requires submission of a new LVE application. 40 C.F.R. § 723.50(i). If EPA fails to take action on an LVE application after twenty-one days, the applicant is free to begin manufacturing or importing.

LVE Listing

EPA publishes in the Federal Register a notice listing each LVE granted. In addition, EPA adds the low volume chemical to a "low volume" list maintained by the Agency. When a new LVE application is submitted, EPA searches this list. If EPA has already granted an LVE for that chemical, the subsequent application for an LVE is denied and the LVE applicant is required to file a PMN before beginning commercial manufacture. EPA does not require the filing of an NOC for an LVE chemical and does not place it on the TSCA Inventory. Hence, LVE chemicals remain "new" chemical substances under § 5.

LVE Revocation

If new information causes the LVE chemical to become ineligible for the LVE, EPA will revoke the exemption. For example, EPA may learn that a company has manufactured more than the permitted 1,000 kilograms, or the Agency may receive information in a TSCA § 8(e) substantial risk notice indicating that the compound in fact may present a substantial risk. Under such circumstances, EPA would no longer be able to make the finding that the LVE chemical will be produced in annual quantities of 1,000 kilograms or less and "will not present an unreasonable risk" as required by § 5(h)(4). Before revoking the LVE, however, EPA will notify the manufacturer or importer in writing of the Agency's intent to revoke.

After receiving notice of EPA's intent to revoke, the manufacturer or importer within fifteen days may file objections or an explanation of its "diligence and good faith" in attempting to comply with the terms of the exemption. 40 C.F.R. § 723.50(g)(2). If the LVE holder was manufacturing, processing, distributing, or using the LVE chemical at the time of the notice and files objections or an explanation, the LVE holder may continue to manufacture the LVE chemical until EPA makes a final determination.

Within fifteen days of receiving the objections or explanation, EPA will make a final determination whether the chemical remains eligible for the LVE. If so, EPA will leave the LVE in effect. If not, then within twenty-four hours of notification by EPA, manufacture must cease -- unless the Agency finds that the manufacturer has acted in good faith to meet the terms of the exemption. Id. If the Agency makes a "good faith" finding, the LVE holder may continue or may resume manufacturing, processing, distributing, or using the LVE chemical so long as the LVE holder submits a PMN within fifteen days of such notification.

EPA created this "good faith" provision to address industry concerns that the immediate cessation of manufacture could disrupt marketing plans and customer relations -- even though EPA had not found an unreasonable risk (but simply could not make the required affirmative finding that the LVE chemical presents no unreasonable risk). Thus, in those instances where a company no longer qualifies for the LVE but has operated in good faith, the Agency permits continued manufacture until the PMN is filed and clears review. Of course, the Agency may determine that the LVE chemical should be regulated under §§ 5(e) or 5(f). If so, then the LVE holder must cease manufacturing, processing, or using the LVE chemical, notwithstanding due diligence and good faith. See 40 C.F.R. § 723.50(g)(2)(B)(vi).

Polymer Exemption

Under TSCA § 5(h)(4), EPA provides for expedited (twenty-one day) review for polymers that are made from a specified list of reactants or have a number-average molecular weight greater than 1,000. 40 C.F.R. § 723.250(e). EPA issued the polymer exemption after determining that the class of polymers eligible under the polymer exemption rule "would significantly limit the risks to human health and the environment that exempt polymers may present." 49 Fed. Reg. 46,084 (1984).

Applicants for the polymer exemption must use the standard PMN form and indicate on the first page of the form that it is a polymer exemption application. The applicant must provide the following information: manufacturer's name, site of manufacture, chemical identity, the number-average molecular weight, residual monomer and other reactants and low molecular weight species content, impurities, production volume, and use. 40 C.F.R. § 723.250(f). As with the LVE application, information on exposure is not required in the polymer exemption application.

As with all exemptions under § 5(h)(4), including the LVE, EPA also must find "no unreasonable risk" in order to grant a polymer exemption. If it makes this finding, EPA will publish a notice in the Federal Register that the twenty-one day review period has expired.

If EPA cannot make this "no unreasonable risk" finding, it will notify the submitter that the substance is ineligible for expedited review. Under those circumstances, EPA automatically will extend the review period to ninety days. For example, if the Agency determines that the polymer should be considered for regulatory action under TSCA §§ 5(e) or 5(f), or that "unresolved issues concerning toxicity or exposure require further review," it will extend the review period. 40 C.F.R. § 723.250(l)(1). If EPA does extend the review period, the manufacturer either must withdraw the polymer exemption application or agree to suspend the review period and within sixty days to submit the additional information necessary to constitute a full PMN. Thereafter, EPA will review the polymer under EPA's standard PMN review procedures. (See section IV.)

Polymer Exemption Requirements

In order to be eligible for the polymer exemption, applicants must meet three criteria: (1) the chemical must be a "polymer" as defined in 40 C.F.R. § 723.250(b)(12); (2) it must not be specifically excluded by 40 C.F.R. § 723.250(d); and (3) it must have a certain number-average molecular weight or be a polyester of a certain type as set forth in 40 C.F.R. § 723.250(e).

Polymer Definition

For purposes of the polymer exemption, EPA defines a polymer as:

[A] chemical substance that consists of at least a simple weight majority of polymer molecules but consists of less than a simple weight majority of molecules with the same molecular weight. Collectively, such polymer molecules must be distributed over a range of molecular weights wherein differences in molecular weight are primarily attributable to differences in the number of internal subunits.

40 C.F.R. § 723.250(b)(11).

A "polymer molecule" must consist of at least four covalently linked "subunits," which in turn must contain at least two "internal subunits." In addition, "[t]he minimum content for 'polymer molecules' in a 'polymer' is 50 weight percent." 49 Fed. Reg. 46,069 (1984). These polymer molecules must be distributed over a range of molecular weights, with differences in the molecular weight attributable primarily to differences in the number of internal subunits and "not solely to changes in the number of pendant groups or similar subunits." Id. Also, the polymer must consist of "less than 50 weight percent of any molecules with the same molecular weight." Id. This definition of polymer encompasses the category of polymeric substances which EPA considers in general do not pose a risk. All polymers must meet this definition to be eligible for the exemption.

Inventory Listing of Polymers

Unlike the situation with an LVE, the manufacturer under a polymer exemption must file an NOC within 30 days after commencing manufacture of the polymer. Then EPA will add the polymer substance to the TSCA Inventory, using the following identifying information: "the name of each reactant used at greater than 2 percent, and those reactants used at less than 2 percent that are identified by the manufacturer to be part of the chemical identity; the maximum weight percent of each monomer or other reactant that will be present as a residual in the polymer[14]; the maximum weight percent of all material below 500 absolute molecular weight and below 1000 absolute molecular weight in any composition that will be manufactured; and for non-polyester polymers, a minimum average molecular weight of 1000." 40 C.F.R. § 723.250(o).

By listing a polymer using its maximum oligomer content, maximum residuals, and the identity of its starting material, EPA limits the polymer to the identity and properties which were reviewed by EPA under the exemption. Thus, any change in the polymer potentially may result in a new chemical substance subject to reporting under TSCA § 5. This rigid chemical identity also affects the applicability of the two percent reporting rule.

Under the two percent reporting rule, some minor structural changes to a polymer are permitted without a new PMN. The use of additional monomers or reactants will not result in a "new chemical substance" subject to the § 5 reporting requirements if each of the additional monomers or reactants amounts to two percent or less of the total molecular weight of the polymer. (For a complete discussion of the two percent reporting rule, see Chapter VI.) Furthermore, the PMN submitter can elect whether or not to identify those reactants present at two percent or less in the chemical name which goes on the Inventory.

By contrast, when a manufacturer applies for a polymer exemption and describes the polymer for the Inventory, thereafter the manufacturer cannot modify the polymer with additional monomers or reactants, even at two percent or less, if the addition causes any of the residual monomers or reactants or the species below 500 or 1000 absolute molecular weight to exceed the maximum percentage by weight that the manufacturer stated in the exemption application. Thus the polymer exemption applicant gains an expedited Agency review, but loses some of the ability to modify the polymer in the future.

Revocation of the Polymer Exemption

If EPA determines that an exempted polymer no longer meets the exemption criteria, the Agency will notify the manufacturer of this finding by telephone and by letter. If a manufacturer is processing, distributing, or using the substance at the time of notification, the manufacturer may continue to do so if he files objections or an explanation within 15 days of receiving written notice. 40 C.F.R. § 723.250(q)(2). Any manufacturers

---

[14]/ Every monomer or other reactant must be included, regardless of its percentage by weight.

not manufacturing at the time of notification, however, cannot commence manufacture until EPA has made a final determination with respect to a polymer's eligibility. Id.

As with the LVE, if EPA determines that the substance no longer meets the polymer exemption criteria but that the manufacturer acted with "due diligence and in good faith," the manufacturer may continue manufacturing, if: (1) it was actually manufacturing, processing, distributing, or using the exempted polymer at the time of telephone notification; and (2) a full PMN is submitted on the new substance within 15 days of telephone notification of ineligibility. 40 C.F.R. § 723.250(q)(2)(i)(B).

Failure to manufacture the polymer substance in accordance with the exemption criteria may subject the manufacturer to an enforcement action. A manufacturer will not be subject to penalties, however, for continuing commercial activity from the date of telephone notification until: (1) the manufacturer objects or explains, for which he has up to 15 days, and (2) the Administrator notifies the manufacturer of his decision, for which he may take up to 15 additional days. 40 C.F.R. § 723.250(q)(2)(i)(C).

OTHER EXEMPTIONS FROM THE PMN REQUIREMENTS

In addition to exempting the low volume chemicals and polymers from the PMN requirements, EPA has used its exemption authority under TSCA § 5(h)(4) to exempt several other types of new chemicals. These exemptions will not be discussed here, but they include: (1) "peel apart" film articles (the "Polaroid exemption"); (2) new chemicals imported in articles; (3) impurities, byproducts, and non-isolated intermediates; (4) chemicals formed incidental to exposure to other chemicals; (5) chemicals formed during the manufacture of an article; (6) chemicals formed incidental to the use of certain additives; and (7) polymer salts.

IV. TSCA § 5: PREMANUFACTURE NOTIFICATION

After determining that a chemical substance is not listed on the Inventory, and not excluded or exempted from the § 5 requirements, the company intending to manufacture or import the chemical for commercial purposes must file a PMN under § 5(a).[15/] The company then must decide whether to file a full PMN or, if the chemical qualifies, a low volume or polymer exemption application. Although the standard PMN form can be used for any of these applications, each has distinct requirements.

DETERMINING THE TYPE OF PMN

If a chemical does not qualify for the low volume or polymer exemptions, the manufacturer must file a PMN. (See section III.) EPA has published an "Instructions Manual for Premanufacture Notification of New Chemical Substances" (instructions manual) to assist submitters in preparing a PMN. The instructions manual is available from the TSCA Assistance Office. In addition, a submitter can call the Prenotice Coordinator with questions on preparing the PMN (or exemption applications).

There are essentially three types of PMNs: (1) full or standard PMN; (2) consolidated PMN; and (3) joint submission.

[15/] If a company intends to manufacture or import a new chemical substance solely for test marketing purposes, it can submit a test marketing exemption (TME) application. (See section III.)

Full PMN

A company prepares a full or standard PMN by filling out a standard PMN form for each new chemical. (The PMN form is published at 40 C.F.R. pt. 720, appendix A, and is available from the TSCA Assistance Office.) In general, EPA intends that the submitter use one PMN form for each new chemical when submitting standard PMNs or exemption applications.

Consolidated PMNs

EPA developed the consolidated PMN in response to requests from submitters who had been filing groups of PMNs for structurally similar chemicals. According to EPA, "[t]hese notices contained a large amount of duplicative information, such as exposure, release, or use data." 48 Fed. Reg. 21,734 (1983). Thus, EPA developed a consolidated PMN to cover two or more new chemicals similar in structure or use. EPA does not intend the consolidated PMN to be used for a series of intermediates and a final product, but rather intends it for chemicals with the same or similar structures and uses. Before filing a consolidated PMN, the Prenotice Coordinator must give approval. A manufacturer should submit a written request for consolidation and identify the chemical substances to be consolidated. The Prenotice Coordinator will grant approval to file a consolidated PMN over the phone, but only for a few chemicals. Where the Prenotice Coordinator and the OTS chemists determine that the chemicals are not sufficiently similar, they will disapprove a consolidation request.

Joint Submissions

In some cases, another person may have some of the information required to complete a PMN, yet be unwilling to reveal that information to the submitter. For example, a foreign supplier may possess the specific chemical identity but refuse to reveal it to the domestic importer who is filing the PMN. In order to accommodate these situations, EPA developed the "joint submission." 40 C.F.R. § 720.40(e)(2).

To file a joint submission, each joint submitter must complete a PMN form, sign the certification statement, and assert any confidential business information (CBI) claims. Each submitter should identify the joint submitter and should indicate which information the other joint submitter will supply. Each submitter is required to complete all mandatory sections of the form to the extent the information "is known to or reasonably ascertainable by the submitter." 40 C.F.R. § 720.40(d). The review period will not begin until EPA has received all of the required information.

Special Cases: Polymer and Low Volume Exemptions

The information requirements for the polymer and low volume exemptions differ slightly from those of a full PMN. In completing a polymer exemption application, submitters need not provide any information on exposure or environmental release. 40 C.F.R. § 723.250(f). The submitter, however, must provide the following certification language:

(A) The notice includes all test data and other required data.

(B) The person [insert company name] submitting the notice intends to manufacture or import the polymer for a commercial purpose other than for research and development.

(C) All information provided in the notice is complete and truthful as of the date of submission.

(D) The new chemical substance meets the definition of polymer, is not specifically excluded from the exemption and meets the conditions of the exemption.

40 C.F.R. § 723.250(f)(2)(xii). This <u>exact</u> language must be used for the polymer exemption application or EPA will declare the application incomplete and will not start the twenty-one-day notice review period running.

Similarly, submitters preparing a low volume exemption application need not provide exposure and environmental release information, but must certify to the following:

(A) The manufacturer intends to manufacture or import the new chemical substance for commercial purposes, other than in small quantities solely for research and development, under the terms of this section.

(B) The manufacturer [substitute your company name] is familiar with the terms of this section and will comply with those terms.

(C) The new chemical substance for which the notice is submitted meets all applicable exemption conditions.

40 C.F.R. § 723.50(e)(vi).

COMPLETING THE PMN FORM

The Submitter's Basic Obligations

The PMN form requires information on the submitter's identity, specific chemical identity, production volume, use, worker, user and consumer exposures, and environmental fate. The submitter must provide this information to "the extent it is known to or reasonably ascertainable by the submitter." 40 C.F.R. § 720.45. In essence, the PMN represents the submitter's best estimates of the future production, distribution, and use of the new substance at the time the notice is filed. Unless EPA issues an order or rule on the PMN under §§ 5(e) or 5(f), the submitter is free to change most of these projections after the PMN clears review (i.e., the submitter is not "locked-in" to the estimates represented in the PMN). The submitter, however, may not change the chemical identity, since this is the only chemical which EPA will consider in reviewing the PMN. The submitter is required to maintain records of the information contained in the notice for five years from the date of the notice of commencement of manufacture. 40 C.F.R. § 720.78.

Premarket Data Requirements

Submitters are not required to develop any minimum amount of data on the new chemical prior to submission of the PMN. Absent an applicable test rule, the submitter need file only those data which he believes demonstrate that the manufacture, processing, distribution in commerce, use, and disposal of the substance will not present an unreasonable risk. 40 C.F.R. § 720.40(h). The submitter, however, must tender any test data in his possession or control related to environmental or health effects of manufacturing, processing, distributing, using, or disposing of the new chemical substance, or any mixture or article containing it. 40 C.F.R. § 720.50. In 1986, EPA revised its definition of "possession or control" to include data in the files of employees engaged in R&D, test marketing or commercial production of the substance and to agents to the extent the files are kept in his capacity as agent. 40 C.F.R. § 720.3(y).

The requirement to submit test data on the new chemical applies to data on the chemical in its pure, technical grade, or formulated form. The submitter must provide a full report or standard literature citation for health effects data, ecological effects data, environmental fate characteristics, physical and chemical properties, exposure, and monitoring data. This applies even to incomplete studies, reports, and tests. If the data do not appear in the open scientific literature, a full report must be provided. 40 C.F.R. § 720.50(a)(1)-(3).

In 1986, the Agency also clarified its position to state that a PMN submitter must furnish agricultural screening data with PMNs. EPA, New Chemical Information Bulletin, Submission of Agricultural Screening Data in PMNs at 2 (Dec. 1986) (available from the TSCA Assistance Office). According to EPA, many companies routinely screen all new chemical substances for potential biological activity. If the company thereafter intends the chemical for a TSCA use, e.g., use as a pesticide intermediate, it must submit any agricultural screening data with the PMN. EPA considers these to be "environmental effects data" under § 5(d)(1)(B). Companies may submit summaries of primary agricultural screening results, rather than detailed descriptions of the test protocol and results. Id. The Agency may request more detailed information on a voluntary basis.

In addition to test data, submitters also must submit any other information and data on the new chemical to the extent that it is "known to or reasonably ascertainable by the submitter." 40 C.F.R. § 720.40(d). EPA does not require submission of exposure data for non-U.S. populations or ecosystems, or submission of efficacy data. 40 C.F.R. § 720.50(d)(2), (3). But EPA does require submission of health and environmental effects data, even though they do involve non-U.S. populations. Id.

If during the review period the submitter comes into possession, control, or knowledge of any "new information that materially adds to, changes, or otherwise makes significantly more complete" the information contained in the PMN, the submitter must give the information to EPA within ten days of its receipt, but in no event less than five days before the end of the review period. 40 C.F.R. § 720.40(f). If the new information becomes available within the last five days of the review period, EPA must be notified immediately.

If a § 4 test rule is promulgated on a substance prior to the submission of a § 5 notice, the submitter must file the test data in the form and manner specified by the TSCA § 4 test rule. Failure to submit test data will render the submission incomplete. If EPA has granted the submitter an exemption under TSCA § 4(c) from the requirement to conduct tests and to submit data, a PMN or exemption application cannot be submitted until EPA receives the test data. 40 C.F.R. § 720.40(g)(2). Until the required data become available, EPA simply will not process the PMN.

If the submitter has received an exemption under § 4(c) and another person previously has submitted test data to EPA, the submitter has two options. He may either submit the test data himself or provide information describing the name of the person who submitted the data, the date on which the data were submitted to EPA, a citation to the test rules, and a description of the exemption and a reference identifying it. 40 C.F.R. § 720.40(g)(3). Any person who intends to manufacture or import a new substance that is on the "risk list" under § 5(b)(4) must file data that demonstrate that the substance will not present an unreasonable risk. 40 C.F.R. § 720.40(h). If the chemical is also subject to a § 4 test rule, however, the submitter either must furnish the data or must qualify for an exemption from doing so under § 4(c), or EPA will deem the PMN incomplete. 40 C.F.R. § 720.40(g).

Practical Tips

Sections of the notice form should not be left blank. If the information is unknown or not applicable, then state as much in those sections. Any "blank" section or part may prompt EPA to consider the notice incomplete. The submitter should check the notice form to make sure all confidential information has been claimed as such. In addition, if any information has been claimed confidential, the submitter must provide EPA a copy with the CBI deleted (the so-called "sanitized" copy) at the same time he files the PMN or exemption application.

PMN FORM PART I: GENERAL INFORMATION

Submitter Identification — Section A

The person who intends to manufacture or import a new chemical substance must submit a PMN. 40 C.F.R. § 720.22. Yet, only manufacturers that are licensed, incorporated, or doing business in the United States may submit a PMN. 40 C.F.R. § 720.22(a)(3). A manufacturer includes the person who contracts with another (toll or contract manufacturer) to make a chemical substance for him. In this case, the person who contracts with the toll manufacturer is considered a manufacturer if he specifies the chemical identity, the total amount of the basic production technology, and the toll manufacturer produces the substance exclusively for him. 40 C.F.R. § 720.22(a)(2). A person who simply orders a non-specific chemical having certain properties, however, is not a manufacturer under this definition and, therefore, is not eligible to file the PMN.

Importers of new chemical substances also are subject to the PMN requirements (unless the substance is excluded or exempted). 40 C.F.R. § 720.22(b). An importer is the person primarily liable for the payment of duties; he can be either the consignor, the importer of record, the actual owner, or any transferee. 40 C.F.R. § 720.3(1). In the case where several parties are involved in the import transaction, the "principal" importer is responsible for submitting the notice. 40 C.F.R. § 720.22(b)(2). The principal importer is the importer who selects the substance to be imported, who knows that the substance actually will be imported rather than manufactured domestically, and who specifies the identity of the substance and the actual amount to be imported. As with a manufacturer, a principal importer must be licensed, incorporated, or doing business in the United States. 40 C.F.R. § 720.3(z).

Agents

Manufacturers or importers required to submit a notice may designate an agent to submit the notice for them. The agent is often a consultant or an attorney. Both the agent and the manufacturer or importer must sign the certification statement. 40 C.F.R. § 720.40(e).

Technical Contact

The submitter must also list its technical contact. The technical contact serves as the company's representative on the PMN. He is generally a person who has actually prepared the PMN or is familiar with the details of the chemical substance. The technical contract should be based in the United States, fluent in English, and technically qualified.

Previous Communications With EPA

The submitter must indicate any prenotice communication, test market exemption, or bona fide request previously made of the Agency.

Chemical Identity — Section B

EPA considers the specific identity of the new chemical substance to be critical information in the PMN. Indeed, if the submitter fails to describe the chemical identity adequately, EPA will declare the PMN incomplete and will not start the review period. Even if a substance is classified as a trade secret under the Occupational Safety and Health Administration (OSHA) Hazard Communication Standard (29 C.F.R. § 1910.1200), the submitter still must report its specific chemical identity to EPA.

EPA classifies chemicals into three categories: Class 1 substances, Class 2 substances, and polymers. 40 C.F.R. § 720.45(a). This classification scheme is the same one that was used in the initial Inventory reporting rules.

Class 1 Substances

Class 1 substances are those whose composition (except for impurities) can be represented by a structural diagram. For Class 1 substances, the submitter must include a specific chemical name using the standard rules of chemical nomenclature, the molecular formula, the Chemical Abstract Services (CAS) registry number (if known), and a structural diagram. 40 C.F.R. § 720.45(a)(1).

Class 2 Substances

Class 2 substances are those whose composition cannot be represented by a structural diagram. Class 2 substances are often derived from natural sources or complex chemical reactions. For Class 2 substances, the submitter must include the specific chemical name, the molecular identity, the immediate precursor compounds, a description of the reaction process, the range of composition and the typical composition, and, if possible, a structural diagram. EPA recommends that the submitter contact the Prenotice Coordinator by letter in order to obtain specific guidance before submitting a PMN on a complex reaction product. 40 C.F.R. § 720.45(a)(2).

Polymers

A polymer is defined as "a substance composed of molecules characterized by the regular or irregular repetition of one or more types of identical monomeric units." EPA, Instructions Manual for Premanufacture Notification of New Chemical Substances at 2 (May 23, 1983) (instructions manual). The Polymer Exemption rule, however, defines a polymer in a more restricted manner. See section III. Submitters intending to manufacture a new polymer must indicate the lowest number-average molecular weight of any composition, and must provide the chemical name and CAS number of each reactant, including those used at two percent or less, the absolute molecular weight below 500 and the typical weight percent of each reactant below 1000, the maximum residual of each monomer present in the polymer, and, if possible, a structural diagram. 40 C.F.R. § 720.45(a)(3).

The Two Percent Reporting Rule For Polymers. On the TSCA Inventory, EPA only required the reporting of monomers and other reactants in a polymer present at two percent or greater. The final PMN rule requires that *all* constituents of a polymer be listed on the PMN, but provides that the submitter may elect which constituents, present at two percent or less, will be incorporated into the Inventory description of the polymer. Id. Thus, by implication, EPA has incorporated the two percent rule into the PMN rule. In addition, EPA discusses the two percent rule in its instructions manual at 3.

In issuing the polymer exemption rule, EPA again clarified the relationship between the two percent rule and the PMN requirement:

> However, § 720.45(a)(3) of the Premanufacture Notification Rules and the PMN form require the reporting of monomers or other reactants used at any weight percent. The submitter may then choose which, if any, of those monomers and reactants used at 2 weight percent or less are to be included as part of the identity of the polymer to be entered on the Inventory.

49 Fed. Reg. 46,076 (1984) (codified at 40 C.F.R. pt. 723).

On occasion, EPA also has provided advice to individual companies about the two percent rule:

> The Inventory rules state that monomers used at two percent (by weight) or less _may_ be included as part of the Inventory description of the polymer (the two percent rule). [Emphasis in original.] For example, if you want to manufacture for commercial purposes a polymer from A and B, each used at greater than two percent, and C, used at two percent or less (by EPA's definition), and if neither category A-B nor category A-B-C are listed on the Inventory, you must submit a premanufacture notice. You may choose whether to include component C as part of the Inventory category description. If you do not include C in the description, and later want to manufacture the polymer using C at greater than two percent, you will have to submit another premanufacture notice. If you want to manufacture the polymer using C at two percent or less, and if polymer category A-B is listed on the Inventory, you do not have to submit a premanufacture notice. You may submit a notice if you want to have category A-B-C listed on the Inventory. However, if only A-B is listed on the Inventory, you must submit a premanufacture notice if you want to manufacture the polymer with C used at greater than two percent.
>
> <u>Although you are not required to include monomers or reactive ingredients used at two percent (by weight) or less for purposes of describing a new substance, you must report information on them in the premanufacture notice for the polymer.</u> [Emphasis added.]

The answers to your specific questions follow.

> 1.a. Polymer category A-B-C-D <u>does</u> include polymer A-B-C-D-E since E is <u>used</u> (charged initially) at less than two percent (by EPA's definition) in the manufacture of the polymer. [Emphasis in original.]
>
> b. Polymer category A-B-C-D <u>does not</u> include polymer A-B-C-D-E if E is <u>used</u> (charged initially) at greater than two percent (by weight). [Emphasis in original.]

Letter from John B. Ritch, Jr., Director, Industry Assistance Office (Mar. 19, 1981) (addressee name deleted).

Free Radical Initiators. One particular area of confusion in the past has been whether to include free radical initiators in the description of a polymer. The Agency has defined the phrase "monomers and other reactants" such that it:

> [A]pplies to those reactive agents that are used intentionally to become chemically part of the polymer composition. These reactive agents include all monomers, e.g., vinyl chloride, acrylamide, terephthalic acid, and any other reactive agents that are intended to be incorporated into the polymer. Reactants other than monomers include crosslinking agents, chain-terminating agents, *free radical initiators*, and any other reactant which is intended to be incorporated into the structure of the polymer. This may include such substances as monohydric alcohols, e.g., methanol, and monofunctional amines, e.g., butyl amine, and acids or bases used to form a salt of the polymer.

Letter from Charles L. Elkins, Director, Office of Toxic Substances (Feb. 9, 1987) (emphasis in original) (addressee name deleted).

Free radical initiators are used in the manufacture of certain classes of polymers. Fragments of these initiators may become chemically incorporated into the polymer as it is manufactured. At one time, EPA took the position that such chemical incorporations did not mean that the initiator necessarily became part of the Inventory description of the polymer. Instead, the Agency asserted that whether or not the initiator became part of the polymer description turned on the intent of the manufacturer in using it:

> The Agency recognizes that "intent" plays an important role in a company's determination of whether or not to include a free radical initiator in the polymer description. Likewise, you reference correspondence from John Ritch, former director of EPA's Industry Assistance Office, dated October 20, 1980, that "seemed to once again leave the question of whether or not to include initiators to 'intent.'" The Agency acknowledges that it is the manufacturer's responsibility to make a good faith effort in determining whether or not a free radical initiator added at a concentration of greater than two weight percent is intended to be chemically incorporated into the polymer. While recognizing the manufacturer's "intent" as the determining factor, the Agency reserves the right to raise questions relating to a reactant role for an initiator when it is used at greater than two weight percent if there is obvious indication of significant incorporation of initiator into a polymer composition.

Id.

EPA eventually recognized that members of the polymer manufacturing industry need more clarification of the Agency's position, and that the "intent" standard was unnecessarily confusing. Accordingly, on June 28, 1989, EPA issued a *Federal Register* notice to clarify the industry obligations. 54 Fed. Reg. 27,174.

The clarification notice states that a polymer that is manufactured with one or more free-radical initiators, in which at least one initiator is used at greater than two percent by weight is considered to be on the Inventory if it falls into one of the following categories: (1) the identical polymer description, without any initiator(s), was on the Inventory as of July 28, 1989; or (2) the identical polymer description, including the same initiator(s) in the name, is on the Inventory. To help future manufacturers of polymers with free-radical initiators determine whether a polymer falls into one of these categories, EPA will flag those polymers that were on the Inventory as of July 28, 1989 which do not have the initiators in the polymer name.

As of July 28, 1989, any polymer manufactured using greater than two percent by weight of an initiator will be subject to PMN requirements unless the polymer is on the Inventory, i.e., unless the polymer meets one of the two criteria given above, or EPA agrees with a claim by the manufacturer or importer that the initiator(s) used at greater than two percent by weight are not incorporated into the polymer structure.

EPA will consider an initiator to be incorporated into the polymer structure if it is known or can be reasonably ascertained that its use primarily involves chemical incorporation of one or more initiator fragments (other than a hydrogen atom) into the polymer. Manufacturers who claim that a free-radical initiator that they use at greater than two percent by weight is not being incorporated into the polymer must provide information to support that assertion. For example, the manufacturer might submit: (1) a description of the activity of the initiator and the extent to which this activity will operate; (2) an explanation of what is intended by this activity role; (3) the extent of incorporation of initiator-derived fragments; (4) any measures taken to prevent or limit incorporation; (5) any product properties or performance factors that would make incorporation undesirable; and (6) any other technical, economic, or utility considerations that would sustain the manufacturer's claims.

Individual Reactants. The two percent rule applies to each individual monomer or other reactant and not to the sum total of all monomers or other reactants. The Inventory Reporting rule, for example, states: "Reporting polymers. (1) To report a polymer a person must list in the description of the polymer composition at least those monomers used at greater than two percent (by weight) in the manufacture of the polymer." 40 C.F.R. § 710.5(c).

Similarly, the PMN rule requires listing of every reactant, but offers the submitter the option of excluding from the Inventory description for the polymer each monomer or other reactant which is used at less than two percent. See 48 Fed. Reg. 41,134 (1983). Thus, the Agency intends that the two percent rule apply to each individual monomer or other reactant rather than to the total of such reactants.

Impurities and By-products

EPA also requires the submitter to list any impurities and by-products as part of the description of chemical identity. An impurity is a substance unintentionally present in the new chemical substance, while a by-product is a separate chemical substance produced without a separate commercial intent. 40 C.F.R. §§ 720.3(d), (m).

Generic Name

The submitter must provide a generic chemical name if the specific chemical identity has been declared confidential. EPA has provided guidelines for developing a generic name in Appendix B of the TSCA Inventory. In addition, the submitter can contact the Prenotice Coordinator for help in developing a generic chemical name. Failure to provide a generic chemical name when the specific chemical name has been declared as CBI will render the notice incomplete.

Synonyms, Trade Name Identification

Finally, the submitter must provide any synonyms and trade names under which the substance has been or will be marketed. 40 C.F.R. § 720.45(c).

Production, Use, and Import Information — Section C

Production Volume

EPA requires the submitter to furnish "reasonable estimates" of the maximum production volume during the first twelve months of production and the maximum twelve-month production during the first three years of production. 40 C.F.R. § 720.45(e). EPA uses these production figures to estimate the potential exposure of workers, users, and consumers and to perform an economic analysis on the chemical. In fact, the production figure (along with the sales price) is a critical factor in determining when a chemical can support the cost of any required testing (either voluntary or pursuant to a § 5(e) order). Production volume is also a crucial factor in the Agency's determination of whether to grant a low volume exemption. By limiting the production to 1,000 kilograms or less per year, the submitter limits potential exposure and hence potential risk to the chemical and establishes his eligibility to seek a low volume exemption.

Use

EPA defines use of a chemical substance in terms of both function and application. EPA has clarified the terms as follows: " 'Function' is related to the inherent physical and chemical properties of the substance (e.g., degreaser, catalyst, plasticizer, ultraviolet absorber). 'Application' refers to the use of the substance in particular processes or products (e.g., a degreaser may be used for cleaning of fabricated metal parts)." Instructions manual at 4. With respect to use, the submitter must estimate the percent of total production volume for each category of use for the first three years (with the percent formulated), and whether the intended use is site-limited, industrial, commercial, or consumer. 40 C.F.R. § 720.45(f).

The submitter must provide a generic use description if a specific category of use has been declared confidential. The generic use description should provide a "sufficient indication of potential exposure." Instructions manual at 5. If a generic use description does not indicate potential exposure, EPA will accept a description of the "degree of containment" (e.g., "contained use, destructive use"). The instructions manual sets forth a list of acceptable containment descriptions. Instructions manual at 5.

Hazard Information

As part of its risk assessment, EPA also will examine any hazard warning statement, label, material safety data sheet, or other information which the submitter intends to provide to any person. Submitters are not required to develop any hazard warning statements, but any that have been developed must be submitted with the notice.

PMN FORM PART II: HUMAN EXPOSURE AND ENVIRONMENTAL RELEASE

EPA requires submitters to provide a description of the manufacturing and processing operations for the new chemical substances both at sites controlled by the submitter and sites controlled by others. EPA utilizes these descriptions in estimating the potential of human and environmental exposure to the new chemical substance.

Industrial Sites Controlled by the Submitter — Section A

The submitter must provide a description of the operation, including the identity of the site where the operation will occur; must indicate whether it is a manufacturing, processing, or use operation and whether it is a continuous run or a batch operation; must indicate the amount and duration of the operation; and must supply a process description. 40 C.F.R. § 720.45(g).

The process description must include a diagram that identifies the major unit operation steps and chemical conversions, the entry point for feedstocks, and the points of release to the environment. The submitter also must provide information on occupational exposure and environmental release and disposal. Even if workers wear protective equipment, the submitter must describe each specific activity during which workers may be exposed along with the physical form of the chemical at the time of potential exposure, and the maximum number of workers potentially exposed.

The submitter also must identify the number of release points to the environment, the amount of the new chemical potentially released at each point, and the medium into which it will be released (air, surface water, ground, etc.).

Industrial Sites Controlled by Others — Section B

To the extent known to or reasonably ascertainable by him, the submitter also must describe processing and use operations controlled by others involved with the new chemical substance. 40 C.F.R. § 720.45(h). The description must include an estimate of the number of sites involved, situations in which exposure to workers or the environment is likely to occur, the percent of the new chemical as formulated in products, and any methods used to control worker exposure and environmental release.

PMN FORM PART III: LIST OF ATTACHMENTS

Finally, the submitter must list any attachments to the notice, such as a material safety data sheet (MSDS), and indicate whether any of them is confidential.

PROTECTING CONFIDENTIAL BUSINESS INFORMATION

TSCA § 14 permits submitters to claim as confidential any information submitted to EPA and also prohibits EPA employees from "knowingly" disclosing this CBI. The claim of confidentiality must be asserted at the time the information is submitted. 40 C.F.R. § 720.80(b).

There are several situations, however, in which EPA can disclose CBI: to any officer or employee of the United States in connection with official duties pertaining to the protection of human health or the environment; to contractors when such disclosure is necessary for their job performance; when the Administrator determines that disclosure is necessary to protect human health and the environment; and when relevant to any proceeding under TSCA. TSCA § 14(a). In addition, TSCA § 14(b) specifically authorizes disclosure of health and safety studies or data obtained from such studies on chemical substances currently in commercial distribution, or subject to the PMN requirements or subject to testing under TSCA § 4. EPA, however, cannot disclose data which reveals manufacturing or production processes or portions of a mixture composed of chemical substances. TSCA § 14(b).

EPA's Handling of CBI

EPA has developed elaborate procedures for handling CBI under TSCA § 14. These procedures include security clearance and training for EPA employees and contractors permitted access to CBI, storage of CBI in secured areas, computer security for CBI, and special methods for creating, transferring, and destroying CBI. See EPA, TSCA Confidential Business Information Security Manual (50 Fed. Reg. 47,108 (1985)).

EPA grants TSCA CBI access separately for each section of TSCA. Before handling TSCA CBI, an EPA employee or contractor must attend a special briefing and must pass a written examination. Those employees and contractors who are authorized access to TSCA CBI are added to an "authorized access list." EPA uses this authorized access list to check whether a particular individual has TSCA CBI access.

The Document Control Officer (DCO) maintains the Document Tracking System in order to track the location and disposition of TSCA CBI. The DCO assigns materials in the Document Tracking System a document control number and a machine-readable bar code which are used to track a document from the time it is received at EPA until it is either transferred to another location or destroyed. All TSCA CBI have a green TSCA CBI cover sheet which clearly identifies the document as CBI. "CBI" is stamped on the first and last pages of the document.

EPA stores TSCA CBI in "secured" areas which are accessible only with the use of a special key card. Persons who do not have authorized access to TSCA CBI can gain entry to a secured area only by signing the visitors log and by being escorted at all times by an authorized person.

EPA has also developed procedures for transferring custody of TSCA CBI. TSCA CBI can be transferred from one authorized person to another only by hand and only if the transferor receives a CBI loan receipt. TSCA CBI can only be transferred to individuals at other facilities by hand or through registered mail, a courier, or the United States Postal Service Express Mail. Regardless of the method, the CBI must be double-wrapped and the outside stamped "To Be Opened by Addressee only."

Asserting CBI in a PMN

To assert a CBI claim, the submitter must check the appropriate box on the PMN from next to the item of information which is CBI. The claims of confidentiality do not have to be substantiated when the PMN form is submitted. A claim of confidential treatment of the specific chemical identity or use, however, must be substantiated when the Notice of Commencement of Manufacture (NOC) is submitted. 40 C.F.R. § 720.85(b).

Any submitter who claims the specific chemical identity and use confidential must provide a generic chemical identity and a generic use description. 40 C.F.R. § 720.87. If any information on the PMN or exemption is declared confidential, the submitter must furnish a copy with all the CBI deleted; this is referred to by EPA as a "sanitized" version of the PMN. 40 C.F.R. § 720.80. TSCA § 5(d)(1) requires that a § 5 notice must be made available to interested persons subject to the confidentiality provisions of TSCA § 14. In addition, EPA must publish a Federal Register notice announcing its receipt of the § 5 notice. Failure to provide a sanitized copy when CBI is claimed will result in EPA declaring the PMN or exemptions application "incomplete." (See above for a discussion of "incomplete" submissions). EPA places this sanitized copy in the public file in the OTS Public Reading Room at EPA Headquarters.

When a NOC is filed, submitters must substantiate claims of confidentiality for chemical identity and use. In order to substantiate a claim, a manufacturer must provide a generic chemical name and answer a series of questions set forth at 40 C.F.R. § 720.85(b). The questions concern the submitter's efforts to protect confidentiality, the harm to the manufacturer's competitive position in the event of disclosure, extent of disclosure to others, whether any other federal agency or court has made any pertinent findings with respect to confidentiality determinations, and the purpose of manufacture or import.

V. REPORTING REQUIREMENTS FOR THE GENERIC SNUR AND SECTION 5(e) CONSENT ORDERS

EPA can impose extensive recordkeeping requirements as part of its negotiating process for § 5(e) Consent Orders and Significant New Use Rules (SNURs). EPA has formalized the recordkeeping requirements in the generic SNUR. 40 C.F.R. § 721. When negotiating a specific SNUR, EPA can choose from, among other items, the recordkeeping requirements discussed below. While EPA has formalized its options for SNURs, similar options are commonly used in § 5(e) Consent Orders.

Hazard Communication Program

Section 721.72 designates as a standardized significant new use any manner or method of manufacture, import, or processing associated with any use without establishing a hazard communication program. When EPA passes an expedited SNUR, the Agency can designate as a significant new use any manufacture, import, or processing associated with any use if the program does not contain the following elements: (1) a written hazard communication program available, upon request, to all employees, contractor employees, and their designated representatives, which, at a minimum, will describe how the requirements of the Generic SNUR Rule for labels, material safety data sheets (MSDSs), and other forms of warning material will be satisfied; (2) a labeling program under which containers of the SNUR substance in the workplace are labeled in accordance with criteria listed in the Generic SNUR Rule; (3) each employer maintains or develops an MSDS for the substance; (4) each employer ensures that employees are provided with information and training on the substance.

If the substance is present in the work area only as a mixture, EPA can set a concentration limit for the substance. An employer is exempt from the provisions of § 721.72 if the concentration of the substance in the mixture does not exceed the set concentration.

An employer need not take additional actions to comply with a SNUR if existing programs or procedures satisfy the hazard communication program requirements. Most of the provisions parallel those of OSHA's Hazard Communication Standard (HCS), but the EPA rule differs in four respects: (1) EPA makes the hazard determination, rather than the employer; (2) EPA provides language for labels and MSDSs, whereas OSHA allows employers to develop language; (3) EPA does not provide trade secret protection as does OSHA's HCS; and (4) EPA requires listing of environmental hazards on labels and MSDSs, whereas OSHA does not. Furthermore, under the SNUR hazard communication program EPA can require labeling, etc., based on a finding that a substance may present a risk. OSHA's hazard communication restrictions generally apply to chemicals which are known to present certain hazards.

Recordkeeping Requirements

Under § 721.125, manufacturers, importers, and processors may be required to keep any of the following records:

o   records documenting dates and volume of manufacture and import of the substance;

o   records documenting volumes of the substance purchased in the United States by processors, dates of purchasers, and names and addresses of suppliers;

- records documenting the names and addresses of all persons outside the site of manufacture, importation or processing to whom the manufacturer, importer or processor directly sells or transfers the substance, transfer dates, and quantities transferred on each date;

- records documenting establishment and implementation of a program for the use of any applicable personal protective equipment required under § 721.63, or the hazard communication program required under § 721.72;

- records documenting the determination required by § 721.63(a)(3) that the chemical protective clothing is impervious to the substance;

- copies of labels and MSDSs required by § 721.72;

- records documenting compliance with any applicable industrial, commercial, and consumer use limitations;

- records documenting compliance with any applicable disposal requirements; and

- records documenting establishment and implementation of procedures that ensure compliance with any applicable water discharge limitation.

VI. BIOTECHNOLOGY REPORTING REQUIREMENTS

EPA's reporting requirements for biotechnology products differ significantly from the reporting requirements for conventional chemicals.

General Guidance for PMNs

In the absence of final regulations addressing biotechnology products, several 1986 and 1987 guidance documents provide useful information on PMN reporting requirements for biotechnology products.[16] Submitters should not use the standard PMN form. The draft guidance advises the submitter to provide EPA with the following data: (1) information on the identity, production, importation, use, human exposure, environmental release and disposal of the microorganism and sites for the activity; (2) all test data and other data related to the health and environmental effects of the new microorganism in the company's possession or control (e.g., risk assessments); (3) test results organized according to the following categories: toxicity, human exposure, environmental exposure, environmental release, and other relevant data; (4) other information relevant to a risk/benefit assessment; and (5) information which has been provided to other federal agencies to obtain regulatory approval of the microorganism.

---

[16] EPA Office of Pesticides and Toxic Substances, Points to Consider in the Preparation and Submission of TSCA Manufacture and Significant New Use Notifications for Microorganisms (July 10, 1986); EPA Office of Pesticides and Toxic Substances, Guidance to Biotechnology Companies Submitting PMNs for Microorganisms to be Released to the Environment (draft) (Oct. 14, 1987); EPA Office of Pesticides and Toxic Substances, Guidance to Biotechnology Companies: Administrative Details for Completing a PMN on Microorganisms (draft) (Oct. 14, 1987).

EPA has developed additional guidelines for submitters who propose to release a microorganism into the environment. The PMN submission should include the following items: (1) a summary which addresses the objectives, significance, and justification of the proposed release; (2) genetic information on the modified organisms to be tested, such as characteristics of the non-modified parental organism and molecular biology of the modified organism; (3) environmental factors relevant to the risks posed by the microorganism, including habitat and geographical distribution; (4) physical and chemical factors which can affect survival, reproduction, dispersal and biological interactions; (5) proposed field trials, including pre-field trial information, conditions at the trial site, containment and monitoring; and (6) risk analysis including the nature of the organism and its function.

On the issue of risk assessment, EPA has provided further guidance. EPA suggests that the submitter should conduct a risk assessment, taking into account: (1) source organisms; (2) the nature of the alteration of source organism genetic material; (3) nature and scope of potential commercial production of the microorganism; (4) use of the microorganism in the environment; (5) potential of the microorganism to survive, multiply and disseminate in the environment; (6) potential contact of the microorganism with other populations; and (7) potential undesirable effects of the microorganism. Although the Agency has developed these informational requirements for the risk assessment, it has not developed a framework for assessing the risks and benefits of microorganisms.

Contents of a PMN for Microorganisms Used in Closed System, Large Scale Fermentations

A July 5, 1989 draft guideline[17]/ outlines the information requirements for microorganisms in closed system, large scale fermentations. The Agency basically requires information that will permit it to evaluate risk and containment mechanisms. The submitter should provide an assessment of the potential hazards posed by the microorganism, based on an evaluation of the characteristics of the host organism, vector, inserted DNA and final construct. The submitter should explain how the manufacturing process involves the organism and how it affects exposure.

The submitter should describe its containment equipment and procedures and explain how its standards fit within the framework of the NIH guidelines on containment. Guidelines for Research Involving Recombinant DNA Molecules, 51 Fed. Reg. 16,958 (1986). The discussion should focus on the means by which the company: (1) minimizes exposure to humans during production; and (2) reduces dissemination of the microorganisms from the plant during and after production. The company's discussion of its containment standards should cover six major areas: (1) transfer of live cultures; (2) inactivation of live cultures; (3) minimization of aerosols; (4) treatment of exhaust gases; (5) sterilization of primary containment equipment; and (6) emergency procedures in case of accidental releases of live organisms. Information should be provided on the effectiveness of containment procedures.

Substantial Risk Information/TSCA § 8(e) Reporting

A 1986 policy statement requires manufacturers, processors and distributors of all biotechnology products subject to TSCA, including those undergoing R&D or involving contained uses, to comply with TSCA § 8(e). 51 Fed. Reg. 23,302. They must notify EPA immediately of any information which reasonably supports the conclusion that the microorganism presents a substantial risk to human health or the environment. EPA

---

[17]/ EPA Office of Pesticides and Toxic Substances, Points to Consider in the Preparation and Submission of TSCA Notification for Microorganisms (Closed System, Fermentation Use) (draft) (July 5, 1989).

-213-

considers the following information to be reportable under § 8(e): "(1) pathogenicity to humans, plants, animals, or microbes, (2) significant ability to displace other organisms in the intended use area, (3) significant potential to transfer genetic material to other organisms, and (4) any other significant potential to cause harm to human health or the environment." Id. at 23,331.

SNUR and TSCA § 8(a) Reporting

In the 1986 policy statement, EPA defined "new," non-agricultural,[18] environmental applications of engineered pathogenic microorganisms, including microorganisms containing pathogenic source genetic materials, as significant new uses. Id. at 23,328-23,330. EPA defined pathogen as "a microorganism that has the ability to cause disease in living organisms" and requested further comment on the definition. Id. at 23,316-317. The § 8(a) rule requests general information on all microorganisms subject to TSCA prior to their release into the environment.

Under the policy statement, however, EPA does not require compliance with either SNUR or § 8(a) reporting. Rather, EPA encourages voluntary compliance with the SNUR requirement ninety days prior to the initial release of the microorganism into the environment. Id. at 23,328-330. EPA has delayed implementation of § 8(a) reporting until a final rule is put into effect.[19]

VII. SECTION 6(a)

EPA has the authority under § 6(a) to impose a variety of controls on chemicals that pose an unreasonable risk, including requirements for recordkeeping. Under § 5(f) EPA has the authority to issue an immediately effective rule under § 6(a). EPA has issued controls, including recordkeeping requirements, in the form of final or proposed rules for metalworking fluids, 49 Fed. Reg. 2772, 24,668 and 36,855 (1984) (codified at 40 C.F.R. § 747); asbestos, 40 C.F.R. § 763 (1989); and hexavalent chromium, 53 Fed. Reg. 10,206 (1988).

VIII. SECTION 6(e): PCBs

Section 6(e) of TSCA directs EPA to regulate the manufacture, importation, processing, distribution in commerce, use, disposal and labeling of PCBs and items containing PCBs. Under this authority, EPA has imposed recordkeeping requirements on various aspects of PCB use.

Storage Requirements

When PCBs and PCB Items (articles or containers) are removed from use, they may be stored for up to one year while awaiting disposal. Certain restrictions apply, however, to such storage. 40 C.F.R. § 761.65. All PCBs and PCB Items must be marked to indicate the

---

[18] USDA has authority to regulate agricultural uses.

[19] Since 1986, EPA has reconsidered its policy on SNUR and § 8(a) reporting. EPA now is considering requiring reporting of all significant new uses but no longer is planning to require any reporting under § 8(a). Until EPA finalizes regulations under TSCA, however, the current Policy Statement remains in effect.

date the PCBs were removed from service. The facility must be properly constructed to contain spills, and the operators must inspect stored PCBs every thirty days, keep records of those inspections, and prepare and maintain on file annual reports. 40 C.F.R. §§ 761.65(b), (c)(5), (c)(9), 761.180.

PCB Cleanup Policy

On April 2, 1987, EPA issued a final policy governing the reporting and cleanup of all spills resulting from the release of materials containing PCBs greater than 50 ppm. 52 Fed. Reg. 10,688 (1987) (codified at 40 C.F.R. §§ 761.120-.135 (subpt. G)). This policy was issued in order to set a uniform cleanup standard among EPA regions, each of which had previously established its own respective cleanup standards, and in recognition that the risks posed by PCBs vary with the amount of PCBs and the location. 52 Fed. Reg. 10,689 (1987).

On October 19, 1988, EPA published amendments and clarifications to the PCB Spill Cleanup Policy, at least in part to ensure that its reporting requirements were consistent with other regulations governing spill reporting. 53 Fed. Reg. 40,882 (to be codified at 40 C.F.R. § 761.125(a)(1).) Under the amended spill policy, any spill that involves a release of more than ten pounds of PCBs _by weight_ must be reported immediately to the appropriate EPA Regional Office. (CERCLA (Superfund) also requires reporting to the National Response Center.)

The Policy classifies PCB spills in the following manner:

o   Low concentration spills (less than 500 ppm PCBs) which involve less than 1 lb. of PCBs by weight (less than 270 gallons of untested mineral oil).

o   High concentration spills (greater than 500 ppm PCBs) or low concentration spills involving 1 lb. or more PCBs by weight (270 gallons or more of untested mineral oil).

52 Fed. Reg. 10,692, 10,693.

PCB Manifesting Rule

The PCB Notification and Manifesting Rule creates four new classifications of parties responsible for "PCB wastes": (1) "generators"; (2) "commercial storers"; (3) "transporters"; and (4) "disposers."

Generators

Generators must comply with the rule's manifesting and recordkeeping requirements, and may have to comply with its notification requirements.

1. _Notification and Identification_. The rule establishes notification and identification requirements for various handlers of PCB waste. Responsible parties (companies or persons) are required to notify EPA on a standard form by April 4, 1990 if they were generating, transporting, storing or disposing of PCB waste prior to the effective date of the rule (February 5, 1990). If they generate or handle PCB wastes after February 5, 1990, they must notify EPA prior to commencing such operations.

The information required on the notification form includes: (1) name of the facility; (2) owner; (3) previous RCRA identification number, if applicable; (4) address of respondent; (5) location of installation; (6) name, title and telephone number of a contact person at the installation; (7) type of PCB activity, i.e., generator, storer, transporter or disposer; and

(8) certification by the owner, operator or an authorized representative that the statements made are true. EPA will assume that the information on the form is not confidential business information unless the submitter provides up-front substantiation of its proprietary nature.

Generators are subject to notification requirements only if they are storers and under the PCB Manifesting Rule a company needs to register only these sites at which it stores PCB wastes. Thus, unlike RCRA (which treats each generation site as a separate "generator") a company can fulfill its notification obligations by registering only its storage sites, even if it generates PCB waste from many other facilities. Furthermore, if several generators, unrelated to each other, share the same storage site and commingle their wastes, they can obtain a single "generator" number. This advantage is somewhat limited, however, because storage facilities at different sites each must submit a unique generator notification. Thus, if a generator owns or operates storage facilities at several sites, each site is treated as a different "generator" of PCB waste.

After notifying EPA, generator/storers will receive a twelve-digit identification number. If a facility already has a RCRA identification number, the same number will be used for TSCA purposes. As of June 4, 1990 it will be illegal to deliver regulated PCB waste without an identification number. All generators that are not storage facilities can use the number "40 CFR PART 761" as their identification number, thereby indicating that they are exempt from the notification requirements.

2. Manifesting. A generator is required to prepare a manifest whenever he "relinquishes control" over waste containing over fifty parts per million ("ppm") of PCBs. Regardless of whether the transport vehicle belongs to the generator or an independent transporter, when waste is placed on a vehicle for off-site transport to another company's facility a manifest is required. If the generator owns or operates the off-site storage or disposal facility, however, no manifest is necessary because the generator has not relinquished control.

EPA has adopted as the PCB waste manifest the same Hazardous Waste Manifest Form utilized under RCRA. The manifest must contain the following information: (1) the manifest document number, including the generator's EPA ID number and a five digit number that the generator adds which is unique to each shipment; (2) name, address and telephone number of the generator; (3) name and ID number of the initial transporter and of any additional transporters, if applicable; (4) the name, site address and ID number of the designated storage or disposal facility; (5) the container number and type; (6) a Department of Transportation description, including shipping name, hazard class and ID number; (7) special handling instructions and additional information including the date of removal from service for disposal; (8) generator's certification; (9) acknowledgement of acceptance by transporter; (10) space for discrepancy reporting; (11) acknowledgement of acceptance by designated facilities; (12) any optional information required by applicable state regulations; and (13) total number of pages in the manifest.

Generators are required to retain the original copy of the manifest until they receive the copy signed by the designated disposal or storage facility. The generator also must retain the signed copies that are by the designated disposal or storage facilities for three years from the date the PCB waste was accepted by the initial transporter. § 761.209.

Generators are subject to tracking requirements above and beyond the completion of the manifesting form. Generators who employ independent contractors to transport their waste must confirm (by telephone or other means) that the designated facility received the waste. Confirmation must be made by the close of business the day after the generator receives the manifest from the designated facility.

If a generator does not receive a copy of the signed manifest from the designated disposal or storage facility within thirty-five days of the date the waste was received by the initial transporter, the generator must contact the facility about the status of the waste. If the generator has not received the manifest within forty-five days of acceptance by the initial transporter, the generator must file an "exception report" with the EPA Regional Administrator for the region in which the waste generation site is located. The exception report includes a copy of the manifest and a signed cover letter detailing the actions taken to reconcile the problem. A generator also must submit an exception report if the generator receives the signed manifest but cannot confirm that the disposer or storer actually received the waste.

Under the existing storage regulations PCBs must be disposed of within one year after they are removed from service. Generators are presumed to meet the requirement if they transfer PCBs to a disposer within nine months of the PCB's removal from service. The new regulations require generators to file a one-year "exception report" when: (1) they transfer PCB waste more than nine months after removal from service; and (2) do not receive a certificate of disposal within thirteen months (the one-year limitation plus the thirty days that a disposer has to mail the certificate of disposal) or receive a certificate of disposal that confirms a date of disposal of more than one year from the date of removal.

3. Recordkeeping. The rule requires generators with (1) at least 45 kilograms (99.4 pounds) of PCBs in PCB containers; (2) one or more PCB transformers; or (3) fifty or more PCB Large High or Low Voltage Capacitors, to maintain annual records and an annual document log for at least three years after the facility ceases using or storing PCBs. The records must include manifests and certificates of disposal. The annual document log will contain the facility's EPA ID number, manifest numbers of PCB waste disposed during the year, and a summary of PCB wastes generated and sent off-site.

Commercial Storers

Commercial storers are required to comply with approval requirements, applicable requirements of the tracking system and recordkeeping requirements.

1. Notification and Identification. Storers and disposal facilities are required to notify EPA of their operations on the standard notification form. EPA will assign each facility an identification number which must be used on manifesting forms.

2. Manifesting. Disposers and storers are required to comply with routine manifesting requirements. Facilities must sign and date each copy of the manifest, return a copy to the generator and maintain their copy until at least three years after the facility closes.

Additionally, storage and disposal facilities must report to EPA discrepancies in quantity or type of waste received. A discrepancy is a variation in weight greater than ten percent or an obvious difference in the type of waste, e.g. a substitution of solids for liquids. If there is a discrepancy the facility must try to reconcile it with the generator or transporter. If it is not resolved within fifteen days of receiving the waste, the facility must send a letter immediately to the Regional Administrator in the Region where its facility is located. The letter describes the discrepancy, the actions taken to reconcile the differences and includes a copy of the manifest.

After April 4, 1990, a storer or disposer receiving unmanifested PCB waste is required to notify EPA and submit a report within fifteen days of receipt of the waste. The report must include information identifying the facility and the waste, if possible; a brief description of why the waste was unmanifested, if known; and the disposition of the waste.

Reports are sent to the Regional Administrator where the PCB waste was generated, if known, and to the Regional Administrator in the region where the receiving facility is located. After receiving the report, the Regional Administrator makes a determination whether the unmanifested waste may be disposed or stored.

Finally, within thirty days of the date of disposal a disposal facility must send the generator a certificate of disposal that identifies the facility and the processes that were used for disposal, and certifies that the wastes were in fact disposed of. A disposal facility must file a one-year exception report when the facility receives PCB waste more than nine months after the PCBs were removed from service and the waste cannot be disposed of within one year of removal from service.

3. <u>Recordkeeping</u>. Commercial storage or disposal facilities must maintain annual records of manifests and certificates of disposal, and a document log summarizing waste disposal. The records must be maintained until at least three years after the facility is no longer used for storage or disposal of PCBs. Chemical waste landfills, however, must maintain records for twenty years after ceasing operations. Beginning July 15, 1991, facilities must also submit an annual report to EPA containing a brief summary of the year's records.

Transporters

Transporters are required to comply with notification and manifesting requirements. After June 14, 1990, transporters cannot transport waste without an ID number or deliver waste to a facility without an ID. Transporters are required to obtain and sign a copy of the manifest if the waste is being shipped off-site; obtain the signature of the party that the waste is being delivered to; and maintain the copy that was signed by the generator for a period of three years from the date the waste was accepted.

IX. IMPORT AND EXPORT REQUIREMENTS

Importer's Certification Requirement

TSCA § 13 requires the Secretary of the Treasury to refuse entry of any chemical substance whether in bulk or as part of a mixture, or any article containing a chemical substance or mixture, into the U.S. customs territory if: (1) it fails to comply with any rule in effect under the Act, or (2) it is offered for entry in violation of (a) § 5 or 6, (b) a rule or order under § 5k or 6, or (c) an order issued in a civil action brought under § 5 or 7. TSCA § 13(a)(1).

Under the authority granted to the Secretary of the Treasury, the U.S. Customs Service issued regulations to implement the provisions of TSCA § 13. 48 Fed. Reg. 34,734 (1983) (codified at 19 C.F.R. §§ 12.118-.127.) These regulations require an importer to certify at the port of entry that either (1) the shipment is subject to TSCA and complies with all applicable rules and orders thereunder or (2) it is not subject to TSCA. 19 C.F.R. § 12.121(a). Customs and EPA refer to the former as a "positive certification" and the latter as a "negative certification". Failure to certify as required under TSCA § 13 may subject an importer to an administrative civil penalty under § 16(a). Between 1986 and 1988, EPA imposed sixty-three civil penalties for failure to certify under § 13 ranging from $500 to $39,600.

In most cases, the Customs entry documentation and TSCA certification will be filed with Customs at the same port at which the shipment is unloaded from the international carrier. It is not uncommon, however, for a shipment to be unloaded at one Customs port and to be transported inland under bond (in-transit bond) to another Customs port where it is entered with Customs. The port of unlading from the international carrier is referred to as the port of arrival. The port where the certification and other entry documentation is filed is referred to as the port of entry. Customs has established approximately ninety ports where entry documents may be filed. They are listed in § 101.3(b) of the Customs Regulations. 19 C.F.R. § 101.3(b).

In some situations, chemicals are merely moved through the United States for export to another destination, such as from Mexico to Canada. Under these circumstances, no certification is required. Also, import certification is not required for entry into foreign trade zones. Customs does require certification, however, for withdrawal if the chemical is intended for United States consumption. EPA, Chemical Imports, Questions and Answers About Requirements Under Section 13 of the Toxic Substances Control Act (Questions and Answers) (from Dec. 14, 1983 meeting) at 23.

EPA has issued a policy statement clarifying how the Agency will interpret and implement chemical substance import requirements. 48 Fed. Reg. 55,464 (1983) (codified at 40 C.F.R. § 707.20). For guidance to assist importers and exporters, see: EPA, A Guide for Chemical Importers/Exporters, Volume 1: Overview (Feb. 1984) and Volume 2: List of Import/Export Chemicals, (August 1986, Addendum July 1, 1986 - April 15, 1987); EPA, Chemicals on Reporting Rules Database (CORR) (Mar. 31, 1989); and Questions and Answers at 23. These documents are available from the TSCA Assistance Office.

Certification Content

The importer or his agent, a Customs broker, can fulfill the importer's obligation to certify compliance with TSCA by filing one of the two following statements, properly signed, with Customs at the port of entry:

[POSITIVE CERTIFICATION]

I certify that all chemical substances in this shipment comply with all applicable rules or orders under TSCA and that I am not offering a chemical substance for entry in violation of TSCA or any applicable rule or order thereunder.

[or

NEGATIVE CERTIFICATION]

I certify that all chemicals in this shipment are not subject to TSCA.

19 C.F.R. § 12.121(a); Questions and Answers at 6.

The importer must use one of the statements as worded; no other language may be substituted. The certification may appear either on the appropriate entry document or commercial invoice, or on an attachment to the entry or invoice. Questions and Answers at 6. The certification may be signed by means of an authorized facsimile signature. 19 C.F.R. § 12.121(c). The Customs Regulations require the importer (or its agent, the Customs broker) to keep a copy of this certification along with other Customs entry documentation for five years. 19 C.F.R. § 162.1a(a)(2), .1b, .1c.

Blanket Certification Procedures

An importer may use a "blanket" certification to cover several shipments of the same chemical made over a one-year period subject to the approval of the appropriate district director. A blanket certification must be on the letterhead of the importer and signed by an authorized official (or imprinted with authorized facsimile signature). Questions and Answers at 7. Blanket certifications need only be filed with each Customs district where importation is anticipated, not at every port of entry within the district. Id. at 9. The commercial invoice for each shipment, however, must contain a reference to the blanket certification as described further below. The district director may revoke a blanket certification "for cause." A copy of each approved blanket certification is sent by the district director to the Cargo Control Program, Office of Inspection and Control, Customs Service Headquarters, Washington, D.C. Id. at 7.

Blanket Certification Content

The following is EPA's suggested format for blanket positive certifications and for related invoice statements:

POSITIVE CERTIFICATION

TO: District Director
U.S. Customs Service

_____, _____
  (city)   (state)

CERTIFICATE

The undersigned, as an authorized officer or agent of (Importer), hereby certifies that all chemical substances in all shipments of product(s) listed herein and imported from suppliers listed below, namely:

PRODUCTS

(list name and HTS [Harmonized Tariff Schedule of the United States] subheading number)

SUPPLIERS

(list name and address)

comply with all applicable rules or orders under TSCA, and (Importer), is not offering a chemical substance for entry in violation of TSCA or any applicable rule or order thereunder.

Dated this __ day of _____, 19__.

_____
(Authorized Signature)

_____
(Title)

Id. at 7.

COMMERCIAL INVOICE STATEMENT
(COMPLYING PRODUCTS)

"Importation of the products described above are subject to certificate on file with the District Director in respect of compliance with TSCA executed by (Importer), on ____, 19_, the terms of which, including the fact of its execution are incorporated herein by this reference."

Id. at 8.

The following is EPA's suggested format for blanket negative certifications and for related invoice statements:

NEGATIVE CERTIFICATION

TO: District Director
U.S. Customs Service
_____, _____
   (city)   (state)

CERTIFICATE

The undersigned, as an authorized officer or agent of (Importer), hereby certifies that all chemical substances in all shipments of product(s) listed herein and imported from suppliers listed below, namely:

PRODUCTS

(list name and HTS [Harmonized Tariff Schedule of the United States] subheading number)

SUPPLIERS

(list name and address)

are not subject to TSCA.

Dated this __ day of _____, 19__.

_____
(Authorized Signature)

_____
(Title)

Id.

COMMERCIAL INVOICE STATEMENT
(PRODUCTS NOT SUBJECT TO TSCA)

"Importation of the products described above are subject to certificate on file with the District Director indicating that they are not subject to TSCA executed by (Importer), on ____ , 19_, the terms of which, including the fact of its execution are incorporated herein by this reference."

Id. at 9.

Export Regulation

TSCA § 12(a) generally exempts from most provisions of the Act any chemical substance, mixture or article manufactured, processed or distributed solely for export from the United States. The recordkeeping and reporting requirements of TSCA § 8, however, continue to apply to such chemical exports. TSCA § 12(a)(1). In order to qualify for the export exemption under § 12(a), the substance, mixture or article must bear a stamp or label stating that it is intended for export. TSCA § 12(a)(1)(B). The export exemption does not apply if the Administrator finds that the chemical substance, mixture or article "will present an unreasonable risk of injury to health within the United States or to the environment of the United States." TSCA § 12(a)(2). In making this determination, moreover, the Administrator may require testing under § 4. Id.

Additionally, § 12(b) requires exporters to notify EPA before shipment abroad of any substance for which test data are required under § 4 or 5(b), regulatory action has been proposed or taken under § 5 or 6, or an action is pending or relief has been granted under § 5 or 7. Thus, as an increasing number of chemicals are subject to § 4 test rules, § 5(e) orders and SNURs, more chemicals will require notification if exported. Upon notification of export of a regulated substance, EPA must notify the government of the importing country that there are data available (in the case of an applicable test rule under § 4 or 5(b)), or that a rule, order, action or relief exists under § 5, 6, 7 or 12(b).

Exporter's Notification Requirement

Pursuant to § 12(b), EPA promulgated its Export Notification Rule, 45 Fed. Reg. 82,844 (1980) (codified as amended at 40 C.F.R. pt. 707, subpt. D). In addition to the export notices required under § 12(b), the rule requires exporters of PCBs or PCB articles to notify the Agency if the PCBs are being exported for purposes other than disposal. 40 C.F.R. § 707.60(c).

Export notification is required regardless of the intended foreign use of the regulated chemical. In other words, in determining whether export notification is required, EPA does not consider it relevant whether the chemical is being exported for use or in a manner that is not regulated domestically under a § 5, 6 or 7 action, rule, or order. 45 Fed. Reg. 82,844, 82,845.

As a matter of policy, EPA has exempted from the export notification requirements regulated chemical substances and mixtures which are exported in articles (except for PCB articles). Thus, export notification need not accompany export of an article, unless the Agency specifically requires export notification for such articles in the context of individual § 5, 6 or 7 rulemakings. 40 C.F.R. § 707.60(b).

The rule currently requires exporters to submit written notice of the first export or intended export to a particular country in a calendar year postmarked the date of export or within seven days of forming the "intent to export" any regulated chemical substance or mixture, whichever is earlier. 40 C.F.R. § 707.65(a)(1)-(3). EPA has proposed to amend the rule to change the current annual notification requirement for chemical substances and mixtures subject to test rules under § 4 to a one-time export notification for each exporter of such chemical for each country of destination. 54 Fed. Reg. 29,524 (1989). Intent to export such regulated substances "must be based on a definite contractual obligation, or an equivalent intra-company agreement, to export the regulated chemical." 40 C.F.R. § 707.65(a)(3).

The export notice must contain the: (a) name and address of the exporter, (b) name of the substance, (c) date(s) of export or intended export, (d) country or countries of import,

-223-

and (e) specific section of TSCA (4, 5, 6, or 7) under which EPA has taken regulatory action. 40 C.F.R. § 707.67(a)-(e). It is significant that exporters are not required to divulge the name of the shipment consignee. In a letter to Congress, EPA acknowledged that § 12(b) does not authorize the Agency to require information on either the foreign customer or the intended use of the exported product. Notices must be marked "Section 12(b) Notice" and filed with the TSCA Document Processing Center, Office of Toxic Substances at EPA Headquarters in Washington, D.C. 40 C.F.R. § 707.65(c).

Within five days of receiving a TSCA § 12(b) export notice, EPA will transmit the following information to the importing country: (1) the name of the regulated chemical; (2) a summary of the regulatory action the Agency has taken; (3) the name of an EPA official to contact for further information; and (4) a copy of the relevant <u>Federal Register</u> notice. 40 C.F.R. § 707.70(a), (b). EPA will send this notice to the country's ambassador in Washington (or other official designated by the foreign government) and to the State Department. 40 C.F.R. § 707.70(c).

X. ASBESTOS HAZARD EMERGENCY RESPONSE ACT OF 1986

Under the Asbestos Hazard Emergency Response Act ("AHERA," TSCA Title II), Pub. L. No. 99-519, § 2, 100 Stat. 2970 (1986), EPA is authorized to implement regulations addressing asbestos exposures in public and commercial buildings. As part of the regulations EPA requires Local Education Agencies (LEAs) to keep records of: (1) response actions and preventive measures employed; (2) fiber release episodes; (3) surveillance activities; and (4) various operations and maintenance activities. 40 C.F.R. § 763.94. In addition, for state and local government workers involved in abatement and disposal, if exposure levels exceed 0.1 fibers/cm^3 a variety of worker protection requirements, including recordkeeping requirements, go into effect. 40 C.F.R. §§ 763.91, 763.120 (1989).

Appendix

Guide to Record Retention Requirements
for
40 CFR Environmental Protection Agency*
and
**29 CFR Dept. of Labor, Occupational
Safety & Health**

***Note:** Attached is a January 1990 Supplement to the Guide to Record Retention Requirements issued in January 1989.

Guide to Record Retention Requirements

in the Code of Federal Regulations

Revised as of January 1, 1989

Published by the Office of the Federal Register National Archives and Records Administration

ENVIRONMENTAL PROTECTION AGENCY

40 CFR

4.9 State agencies participating in relocation assistance program.

To maintain adequate records of its acquisition and displacement activities in sufficient detail to demonstrate compliance with this Part 4.

Retention period: 3 years after each owner of a property and each person displaced from the property receives final payment.

7.85 Recipients of EPA assistance in the operation of programs or activities receiving such assistance beginning February 13, 1984.

To keep nondiscrimination compliance information.

Retention period: 3 years after completion of project or until complaint is resolved when any complaint or other action for alleged failure to comply with nondiscrimination provision is brought before the three year period ends.

30.500 Persons awarded EPA grants, and contractors for all subagreements in excess of $10,000.

Grantee shall maintain books, records, documents, and other evidence and accounting procedures necessary to show (a) amount, receipt, and disposition of all assistance received for project, including non-Federal share, and (b) total costs of the project. Contractors of grantees shall maintain books, documents, papers, and records which are pertinent to specific EPA grants.

Retention period: 3 years except that (1) if any litigation, claim, or audit is started before the expiration of the 3-year period, the records shall be retained until all litigations, claims, or audit findings involving the records have been resolved, (2) records for nonexpendable property acquired with Federal funds shall be retained for 3 years after its final disposition, and (3) when records are transferred to or maintained by EPA, 3-year retention requirement is not applicable to the grantee. The 3-year retention period starts (i) from the date of submission of the final financial status report for project grants, or, for grants which are awarded annually, from the date of the submission of the annual financial status report, (ii) from the date of approval of the final payment request for the last project of a construction grant for WWT works, and (iii) for such longer period, if any, as is required by applicable statute or lawful requirement, or:

(i) If a grant is terminated completely or partially, the records relating to the work terminated shall be preserved and made available for a period of 3 years from the date of any resulting final termination settlement.

(ii) Records which relate to (a) appeals under the Subpart L—Disputes, of this Part, (b) litigation on the settlement of claims arising out of the performance of the project for which a grant was awarded, or (c) costs and expenses of the project to which exception has been taken by EPA or any of it duly authorized representatives, shall be retained until any appeals, litigation, claims, or exceptions have been finally resolved.

Retention period: 3 years.

30.800 Persons awarded EPA grants, and contractors for all subagreements in excess of $10,000.

See 30.500.

31.42 State and local governments receiving Federal grants and cooperative agreements. [Added]

See 7 CFR 3016.42.

35.929-3 Persons awarded grants to assist in the construction of waste treatment works in compliance with the Clean Water Act.

To maintain records as are necessary to document compliance with regulations upon approval and implementing the user charge system.

Retention period: Not specified.

Guide to Record Retention 1989

35.4105 Recipients of technical assistance grants under the Superfund Program. [Added]

(a) To keep and preserve full written financial records accurately disclosing the amount and disposition of any funds; whether in cash or in-kind, applied to the technical assistance grant project.

Retention period: 3 years from date of final Financial Status Report or until any audit, litigation, cost-recovery and/or any disputes initiated before the end of the 3-year retention period are settled, whichever is longer.

(b) The recipient shall require its contractor(s) to keep such full written financial records to adequately establish compliance with terms and conditions of the subagreement.

Retention period: 3 years from closeout of the subagreement unless audit, litigation, cost-recovery and/or any disputes are initiated before the end of the 3 year retention period.

39.115 Persons applying for a loan guarantee for construction of treatment work.

(a) To maintain financial reports and records necessary to reflect the planned and actual receipt of revenue for repayment.

(b) To keep accurate books, records, and accounts relating to the loan, the loan guarantee, and the funds and accounts used to pay the amounts due on the loan.

Retention period: (a) 3 years; (b) not specified.

51.19 Owners and operators of stationary sources emitting air pollutants for which a national standard is in effect.

To maintain records of nature and amount of emission, air sampling data, and other information deemed necessary to determine compliance with applicable emission limitations or other control measures. (State Implementation Plans)

Retention period: 2 years.

51.102 States conducting public hearings on air pollution control implementation plans.

To maintain a record of each hearing. The record must contain, at a minimum, a list of witnesses together with the text of each presentation.

Retention period: Not specified.

Part 52 Owners and operators of stationary sources emitting air pollutants for which a national standard is in effect.

See 51.19 and specific State plans.

53.9 Applicants offering analyzers for sale as ambient air monitoring or equivalent methods.

To maintain an accurate and current list of the names and mailing addresses of all ultimate purchasers of such analyzer.

Retention period: 7 years.

57.404 Primary nonferrous smelter owners.

To maintain records of the air quality measurements made, meterological information acquired, emission curtailment ordered (including the identity of the persons making such decisions), and calibration and maintenance performed on SCS (supplementary control system) during the operation of the SCS.

Retention period: Duration of the NSO.

60.7 Owners or operators of any building, structure, facility, or installation emitting air pollutants.

To maintain records of the occurrence and duration of any startup, shutdown, or malfunction in operation of any affected facility, any malfunction of the air pollution control equipment, or any periods during which a continuous monitoring system or device is inoperative; a file of all measurements, including monitoring and performance testing measurements; and any other records which may be required by applicable subparts.

Retention period: 2 years.

60.23 States adopting plans for control of designated facilities.

To maintain a record of each public hearing for inspection by any interested party.

Retention period: 2 years.

Environmental Protection Agency

60.49b Owners or operators of industrial-commercial-institutional steam generating facilities.

(a) If monitoring of steam generating unit operating condition plan is approved, to maintain records of predicted nitrogen oxide emission rates and the monitored operating conditions, including steam generating unit load, identified in the plan

(b) To maintain records of the amounts of all fuels fired during each day and calculate the annual capacity factor individually for coal, distillate oil, residual oil, natural gas, wood, and municipal-type solid waste for each calendar year.

(c) To maintain records of the nitrogen content of the oil combusted in the affected facility and calculate the average fuel nitrogen content on a per calendar quarter basis.

(d) To maintain records of opacity for facilities subject to the opacity standard under 40 CFR 60.43b.

(e) To maintain records on the calendar date, the average hourly nitrogen oxides emission rates measured or predicted and other information as specified in section cited for each steam generating unit operating day for facilities subject to nitrogen oxide standards under 40 CFR 60.44(b).

Retention period: 2 years following date of record.

60.63 Owners or operators of Portland cement plants. [Added]

To maintain records of visible emissions.

Retention period: 2 years.

60.113a Owners or operators of volatile organic liquid storage vessels (including petroleum liquid storage vessels).

To maintain records of each gap measurement at the plant.

Retention period: At least 2 years following the date of measurement.

60.115b Owners or operators of volatile organic liquid storage vessels (including petroleum liquid storage vessels.)

(a) To maintain a record of each inspection performed identifying the storage vessel on which the inspection was performed; the date the vessel was inspected; and the observed condition of each component of the control equipment (seals, internal floating roof, and fittings).

Retention period: At least 2 years.

(b) To keep a record of each gap measurement performed as required by 40 CFR 60.113(b).

Retention period: 2 years.

(c) After installing control equipment in accordance with 40 CFR 60.112b(a)(3) or (b)(1) (closed vent system and control device other than a flare), to keep a copy of the operating plan and record of the measured values of the parameters monitored in accordance with 40 CFR 60.113b(c)(2).

60.116b Owners or operators of volatile organic liquid storage vessels (including petroleum liquid storage vessels).

(a) To keep readily accessible records showing the dimension of the storage vessel and an analysis showing the capacity of the storage vessel.

Retention period: For the life of the source.

(b) To maintain a record of the VOL stored, the period of storage, and the maximum true vapor pressure of that VOL during the respective storage period.

Retention period: At least 2 years.

60.153 Owners or operators of sewage treatment plants. [Added]

(a) To maintain for incinerators equipped with a wet scrubbing device, records of the measured pressure drop of the gas flow through wet scrubbing.

(b) To maintain records of the measured oxygen content of the incinerator exhaust gas.

(c) To maintain records of the rate of sludge charged to the incinerator, the measured temperatures of the incinerator, the fuel flow to the incinerator, and the total solids and volatile solids content of the sludge charged to the incinerator.

Retention period: 2 years.

60.274a Owners or operators of steel plants that produce carbon, alloy, or specialty steels: Electric arc furnaces, argon-oxygen decarburization vessels, and dust-handling systems.

To maintain data on monitoring, measurement, and monthly operational status.

60.274a

-227-

Retention period: 2 years.

60.275a Owners or operators of steel plants that produce carbon, alloy, or specialty steels: Electric arc furnaces, argon-oxygen decarburization vessels, and dust-handling systems.

During performance testing, to maintain records of any 6 minute average that is in excess of the emission limit.

Retention period: Not specified.

60.310 Owners or operators of affected facilities in metal furniture surface coating operations in which organic coatings are applied.

To keep purchase or inventory records and other data necessary to substantiate annual coating usage.

Retention period: 2 years.

60.343 Owners or operators of lime manufacturing plants.

To maintain records of any 6-month average that is in excess of the emissions specified in 40 CFR 60.342.

Retention period: Not specified.

60.344 Owners or operators of affected facilities of lime manufacturing plants.

See 60.343.

60.434 Owners or operators of affected facilities using waterborne ink systems or solvent-borne ink systems with solvent recovery system.

To maintain records of all coatings used, the results of the reference test method or the manufacturer's formulation data used for determining the VOC content of those coatings.

(b) To maintain records of the amount of solvent applied in the coating at each affected facility.

(c) To maintain records of the amount of solvent recovered by the monitoring device.

(d) To maintain records of the amount of solvent applied in the coating at the facility.

(e) To maintain records of the measurements required by sections 60.443 and 60.445.

Retention period: (a)-(c) 1 month; (d) 12 months; and (e) 2 years.

60.455 Owners or operators of affected facilities in surface coating operation in large appliance surface coating operations.

(a) To maintain records of all data and calculations used to determine VOC emissions from each affected facility.

(b) Where compliance is achieved through the use of thermal incineration, to maintain records of the incinerator combustion chamber temperature.

(c) If catalytic incineration is used, to maintain source daily records of the gas temperature, both upstream and downstream of the incinerator catalyst bed.

(d) Where compliance is achieved through the use of a solvent recovery system, to maintain the source daily records of the amount of solvent recovered by the system for each affected facility.

Retention period: Not specified.

60.465 Owners or operators of affected facilities in metal surface coating operations.

(a) To maintain records of all data and calculations used to determine monthly VOC emissions from each affected facility and to determine the monthly emission limit, where applicable.

Retention period: 2 years.

(b) Where compliance is achieved through the use of thermal incineration, to maintain source daily records of the gas temperature, both upstream and downstream of the incinerator catalyst bed.

Retention period: Not specified.

Retention period: At least 2 years.

(c) To keep records of all replacements or additions of components performed on an existing vapor processing system.

Retention period: At least 3 years.

60.537 Accredited laboratories testing new residential wood heaters. [Added]

To maintain records of all documentation pertaining to each certification test, including the full test report and raw data sheets, technician notes, calculations, and tests results for test runs.

Retention period: For at least 5 years.

60.537 Commercial owners who sell used residential wood heaters (stoves). [Added]

To maintain records of the names and addresses of the previous owners.

Retention period: At least 5 years.

60.537 Manufacturers of new residential wood heaters. [Added]

To maintain records of certification testing data, quality assurance (QA) program results, production volumes and information needed to support a request for a waiver or exemption.

Retention period: For at least 5 years.

60.545 Owners or operators of undertread cementing operations, sidewall cementing operations, green tire spraying operations where organic solvent-sprays are used or Michelin-B operations that use carbon adsorbers.

To maintain continuous records of the organic concentration level of the carbon bed exhaust.

Retention period: 2 years.

60.545 Owners or operators of affected facilities that use catalytic incinerators in the rubber tire manufacturing industries.

To maintain continuous records of the temperature before and after the catalyst bed for catalytic incinerators.

Retention period: 2 years.

60.473 Owners or operators of affected facilities in asphalt processing and asphalt roofing manufacture operations.

To maintain a file of the monitoring results of the temperature of the gas at the inlet of the control device.

Retention period: 2 years.

60.486 Owners or operators of affected facilities in the VOC in the synthetic organic chemicals manufacturing industries.

(a) To keep records of equipment leaks (equipment identification information, dates of leaks, and repair methods.

(b) To keep records on design requirements for closed vent systems and control devices and equipments and valves.

Retention period: 2 years.

60.495 Owners or operators of affected facilities in the beverage can surface coating industries.

(a) To maintain records of all data and calculations used to determine VOC emissions from each affected facility in the initial and monthly performance tests.

Retention period: 2 years.

(b) To maintain daily records of the incinerator combustion chamber temperature where compliance is achieved through the use of thermal incineration.

(c) To maintain daily records of the gas temperature, both upstream and downstream of the incinerator catalyst bed if catalytic incineration is used.

(d) To maintain daily records of the amount of solvent recovered by the system for each affected facility where compliance is achieved through the use of a solvent recovery system.

Retention period: Not specified.

60.505 Owners or operators of bulk gasoline terminals.

(a) To maintain tank truck vapor tightness documentation.

Retention period: Permanent.

(b) To maintain records of each monthly leak inspection and to keep documentation of all notifications to owners or operators of each nonvaportight gasoline tank truck loaded at the affected facility.

Guide to Record Retention 1989

60.545 Owners or operators of affected facilities that use thermal incinerators in the rubber tire manufacturing industries.

To maintain continuous records of the thermal incinerator combustion temperature.
Retention period: 2 years.

60.545 Owners or operators of undertread cementing operations, sidewall cementing operations, green tires spraying operations where organic solvent-based sprays are used, Michelin-A operations, Michelin-B operations, or Michelin-C automatic operations seeking to comply with specified kg/mo uncontrolled VOC use limit.

To maintain records of monthly VOC use and the number of days in each compliance period.
Retention period: 2 years.

60.625 Owners or operators of affected facilities in the petroleum dry cleaner operations.

To maintain records of the performance tests for measuring the flow rate of recovery solvent.
Retention period: Not specified.

60.697 Owners or operators of petroleum refinery wastewater systems. [Added]

To maintain records of: (a) the design and operating specifications for all equipment used to comply to applicable standards in a readily accessible locations; (b) each inspection where a water seal is dry or breached, a cap or plug is out of place, emissions are detected, or a problem is identified, including information about the repairs or corrective action taken; (c) for facilities using a thermal incinerator, continuous records must be maintained of the temperature of the gas stream in the combustion zone of the incinerator. Also, to maintain records of all 3-hour periods during which the average temperature of the gas stream in the combustion zone of the thermal incinerator is more than 28 degree C (50 degree F) below the temperature; and other such records as specified in cited section.
Retention period: 2 years unless otherwise noted.

60.714 Owners or operators of affected coating operations or affected coating mix preparation equipments. [Added]

To maintain records of the monthly weighted average mass of VOC contained in the coating per volume of coating solids applied to each coating.
Retention period: 2 years following the date of the measurements and calculations.

60.724 Owners or operators of facilities that surface coat plastic parts for business machines. [Added]

To maintain at the source, records of all data and calculations used to determine monthly VOC emissions from each coating.
Retention period: For a period of at least 2 years.

Part 60, Appendix F Owners or operators of any building, structure, facility, or installation emitting air pollutants.

To retain records of all measurements from the CEM as required by 40 CFR 60.7.
Retention period: At least 2 years.

61.14 Owners or operators of any stationary sources emitting hazardous pollutants for which a national standard is in effect.

To maintain records monitoring data, monitoring system calibration checks, and the occurrence and duration of any period during which the monitoring system is malfunctioning or inoperative.
Retention period: 2 years.

61.26 Owners or operators of underground uranium mines.

To maintain records of bulkhead inspections and tests (include bulkhead identification number and location and date of each inspection or test).
Retention period: 2 years.

Environmental Protection Agency

61.33 Owners or operators of any stationary sources emitting hazardous pollutants for which a national standard is in effect.

To maintain records of emission test results from stack sampling needed to determine total emissions, as specified in the sections cited.
Retention period: 2 years.

61.34 Owners or operators of any stationary sources emitting hazardous pollutants for which a national standard is in effect.

To maintain records of concentrations at all sampling sites and other data needed to determine such concentrations.
Retention period: 2 years.

61.43 Owners or operators of rocket motor test sites.

To maintain records of air sampling test results and other data needed to determine integrated intermittent concentrating.
Retention period: 2 years.

61.44 Owners or operators of rocket motor test sites.

To retain records of beryllium emission test results and other data needed to determine total emissions.
Retention period: 2 years.

61.53 Owners or operators of any stationary sources which process mercury ore to recover mercury, use mercury chloralkali cells to produce chlorine gas and alkali metal hydroxide, and incinerate or dry wastewater treatment plant sludge.

To maintain records of emission test results and other data needed to determine total emissions.
Retention period: 2 years.

61.55 Owners or operators of mercury-cell chlor-alkali plants.

(a) To maintain records of the results of the emission monitoring.
Retention period: 2 years.
(b) To maintain at the chlor-alkali plant records of the certifications and calibrations.
Retention period: Certification-For as long as the device is used for this purpose. Calibration- For a minimum of 2 years.

To maintain daily records of all leaks and spills of mercury.

61.67 Owners or operators of plants which produce vinyl chloride.

To maintain records of emission test results and other data needed to determine average emissions.
Retention period: 3 years.

61.70 Owners or operators of plants which produce vinyl chloride.

To maintain records of all data needed to determine average emissions.
Retention period: 3 years.

61.71 Owners or operators of plants which produce vinyl chloride.

To maintain records of: (a) The leaks detected by the vinyl chloride monitoring system; (b) leaks detected during routine monitoring with the portable hydrocarbon detector and the action taken to repair the leaks; and (c) emission monitoring. To also keep a daily operating record for each polyvinyl chloride reactor, including pressures and temperatures.
Retention period: 3 years.

61.123 Owners and operators of calciners and nodulizing kilns at elemental phosphorous plants.

To maintain records of emission test results and other data needed to determine total emissions.
Retention period: 2 years.

61.165 Owners or operators of glass melting furnaces which use commercial arsenic as raw materials.

To maintain records to meet the emission limit requirements.
Retention period: 2 years.

61.176 Owners or operators of copper converters.

To maintain records of the visual inspections, maintenance, and repairs performed on each secondary hood systems.
Retention period: 2 years.

Guide to Record Retention 1989

61.185 Owners or operators of arsenic trioxide and metal arsenic production facilities.

To maintain records of all measurements, maintenance and repairs made to the continuo us monitoring system or monitoring device, ambient concentrations at all sampling sites, other data needed to determine such concentrations and other information as specified in section cited.
Retention period: 2 years.

62.4622 Owners or operators of stationary sources emitting air pollutants for which a national standard is in effect.

To maintain records of the nature and amounts of emissions from such source and any other information as may be deemed necessary to determine whether such source is in compliance with applicable emission limitations or other control measures that are part of the plan.
Retention period: 2 years.

80.7 Refiners, distributors, and retailers of gasoline.

To maintain information on bulk shipments and annual gallonage sales of unleaded gasoline as specified in section cited.
Retention period: 6 months.

82.13 Importers of certain chlorofluorocarbons (CFC's) and brominated compounds (halons) to reduce the risk of stratospheric ozone depletion. [Added]

To maintain records on (a) the quantity of each controlled substances imported, either alone or in mixtures; (b) the quantity of each controlled substances were imported; (c) the port of entry through which the controlled substances passed; (d) the country from which the imported controlled substances were imported; the port of exit; and other information as specified in cited section.
Retention period: 3 years.

82.13 Producers of certain chlorofluorocarbons (CFC's and brominated compounds (halons) to reduce the risk of stratospheric ozone depletion. [Added]

To maintain (a) dated records of the quantity of each of the controlled substances produced at each facility; (b) dated records of the quantity of controlled substances used as feedstocks in the manufacture of controlled and in the manufacture of non-controlled substances introduced into the production process of new controlled substances at each facility; (c) dated records of the quantity of ACFC-22 and CFC-116 produced within each facility also producing controlled substances; and (d) other records as specified in cited section.
Retention period: 3 years.

85.1507 Certificate holders importing nonconforming motor vehicles and motor vehicle engines into the U.S.

To maintain adequately organized and index records, correspondence and other documents relating to the certification, modification, test, purchase, sale, storage, registration and importation of that vehicle or engine, including but not limited to specified information required in section cited.
Retention period: 6 years from the date of entry of a nonconforming vehicle or engine imported by the certificate holder.

85.1806 Manufacturers of new motor vehicles or new motor vehicle engines who have been notified that such vehicles or engines are not in conformity with applicable emission standards and regulations.

To maintain records to permit the analysis of recall campaigns as specified in the section cited.
Retention period: 5 years.

85.1904 Manufacturers of motor vehicles or motor vehicle engines who have initiated voluntary emissions recalls.

To maintain records relating to notifications and remedial repairs.
Retention period: Not specified.

85.1906 Manufacturers of new motor vehicles or new motor vehicle engines subject to defect reporting requirements.

To maintain information gathered by the manufacturer to compile emissions defect information reports and voluntary emissions recall reports.
Retention period: 5 years from date of manufacture of the affected vehicles.

Environmental Protection Agency

86.078-7 Manufacturers of new motor vehicles or new motor vehicle engines subject to air pollution control regulations.

To maintain general and specific records including routine emission test records relating to such vehicles as specified in the section cited.
Retention period: 6 years after issuance of all related certificates of conformity; routine emission test records—1 year after issuance of all certificates of conformity to which they relate.

86.440-78 Manufacturers of new gasoline fueled motorcycles subject to emission control standards.

To maintain general and specific records relating to such vehicles as specified in section cited.
Retention period: 6 years; routine emission test records—1 year.

86.605 Manufacturers of new gasoline fueled and diesel light-duty vehicles and new gasoline-fueled and diesel light-duty trucks subject to selective enforcement auditing procedures required by air pollution control regulations.

To maintain general and individual records relating to vehicle emission tests performed pursuant to test orders as specified in the section cited.
Retention period: 1 year after completion of tests.

86.1005-84 Manufacturers of new gasoline-fueled and diesel heavy-duty vehicles and new gasoline-fueled and diesel heavy-duty trucks subject to selective enforcement auditing procedures required by air pollution control regulations.
Retention period: 1 year after completion of tests.

To maintain general and individual records relating to vehicle emission tests performed pursuant to test orders as specified in the section cited.
Retention period: 1 year after completion of tests.

104.25 Manufacturers, Importers and processors of 11-aminoun-decanoic acid.

To retain documentation of information contained in significant new use reports.
Retention period: 5 years from date of submission of the report.

112.7 Owners and operators of onshore or offshore facilities engaged in oil activities.

To maintain written procedures developed for prevention of oil pollution and record of inspection required in 40 CFR Part 112.
Retention period: 3 years.

122.21 Persons holding permits to discharge wastes pursuant to the national pollutant discharge elimination program.

To maintain records of all information resulting from monitoring activities as indicated in section cited.
Retention period: 3 years or longer during period of unresolved litigation or when requested by Director or Regional Administrator.

122.41 Persons holding permits to discharge wastes pursuant to the national pollutant discharge elimination program.

To maintain records of all monitoring information, including all calibration and maintenance records and all original strip chart recordings for continuous monitoring instrumentation, copies of all reports required by the permit, and records of all data used to complete application.
Retention period: 3 years from date of the sample, measurement, report or application.

123.43 State agencies administering national pollutant discharge elimination system permit programs.

To maintain records and information as the Administrator of EPA may reasonably require to ascertain whether the State program complies with the requirements of the Clean Water Act.
Retention period: Not specified.

-230-

Guide to Record Retention 1989

Environmental Protection Agency

125.26 Dischargers submitting applications for a compliance extension for facilities installing innovative technology under the national pollutant discharge elimination program.

To keep records of all data used to complete the request for a compliance extension.

Retention period: For the life of the permit containing the compliance extension.

Part 136, App. A Laboratories performing tests for the organic chemical analysis of municipal and industrial wastewater.

To maintain performance records to document the quality of data that is generated.

Retention period: Not specified.

141.33 Owners or operators of public water systems.

To maintain records of (a) bacteriological analyses; (b) chemical analyses; (c) actions taken to correct violations of primary drinking water regulations; (d) copies of written reports, summaries or communications relating to sanitary surveys of the system; and (e) records concerning variances or exemptions granted to the system.

Retention period: (a) 5 years; (b) 10 years; (c) 3 years after last action taken for each violation; (d) 10 years; and (e) 5 years after expiration of variance or exemption.

142.14 State agencies having primary enforcement responsibilities over public water.

To maintain records of: (a) Microbiological analysis and turbidity analysis; (b) analysis for other contaminants in public water supplies; (c) Inventories of public water systems; (d) sanitary surveys; (e) State approvals; (f) enforcement actions; (g) variances and exemptions issued.

Retention period: (a) 1 year; (b) and (c) 40 years; records retained for at least 10 years, may be transferred to EPA to satisfy the remainder of the required 40-year retention period; (d), (e), and (f) 10 years; (g) 5 years following expiration.

144.28 Owners or operators of Class I, II, and III wells authorized by underground injection control program.

To maintain records of all monitoring information, including all calibration and maintenance records and all original strip chart recordings for continuous monitoring instrumentation, and copies of all reports required by this permit.

Retention period: 3 years. This period may be extended by request of Director at any time.

144.31 Persons applying for permit to operate underground injection wells.

To keep all data used to complete permit applications and any supplemental information.

Retention period: 3 years.

144.51 Persons holding underground injection control permits.

To retain records of all monitoring information, including all calibration and maintenance records and all original strip chart recordings for continuous monitoring instrumentation, copies of all reports required by this permit, and records of all data used to complete the application for the permit.

Retention period: 3 years. This period may be extended by request the State Director at any time.

146.13 Owners and operators of underground wells disposing of fluids.

To retain records of all monitoring information, including all calibration and maintenance records and all original strip chart recordings for continuous monitoring instrumentation, copies of all reports required by this permit, and records of all data used to complete the application for the permit.

Retention period: Not specified.

146.23 Owners and operators of underground wells disposing of fluids.

See 146.13.

146.33 Owners and operators of underground wells disposing of fluids.

See 146.13.

146.72 Owners or operators of Class I hazardous waste wells. [Added]

To retain records reflecting the nature, composition, and volume of all injected fluids.

Retention period: 3 years following well closure.

147.2913 Owners and operators of Class II Injection wells located on the Osage Mineral Reserve, Oklahoma.

To maintain monitoring records on the injection pressure and rate.

Retention period: 3 years or 3 years after enforcement action has been resolved.

147.2922 Owners/operators of Class II injection wells authorized by permit.

To retain all monitoring records on injection pressure and rate.

Retention period: 3 years or if enforcement action is pending, 3 years after enforcement has been resolved.

157.36 Registrants of pesticide products required to be in child-resistant packagings.

To maintain records on description of the packages, copies of certification statements required by section 157.34, and other information as specified in the section.

Retention period: As long as the child-resistant packaging is in effect.

160.29 Testing facilities conducting studies that support applications for research or marketing permits for pesticides regulated by EPA.

To maintain a current summary of training and experience and job description for each individual engaged in or supervising the conduct of a study.

Retention period: 5 years.

160.35 Testing facility quality assurance units conducting studies that support applications for research or marketing permits for pesticide products regulated by EPA.

To maintain (a) a copy of a master schedule sheet of all studies conducted; (b) copies of all protocols pertaining to all studies; and (c) written and properly signed records of each periodic inspection.

Retention period: (a) and (b) 5 years; (c) 2 years.

160.63 Testing facilities conducting studies that support applications for research or marketing permits for pesticides regulated by EPA.

To maintain written records of all inspection, maintenance, testing, calibrating, and/or standardizing operations. Also to maintain written records of nonroutine repairs performed on equipment as a result of failure and malfunction.

Retention period: 2 years.

166.6 State agencies using or applying pesticides pursuant to a quarantine public health exemption.

To maintain records of all such treatments specifying records as contained in cited section.

Retention period: Not specified.

169.2 Producers of pesticides, devices, or active ingredients used in producing pesticides subject to the Federal Insecticide, Fungicide, and Rodenticide Act, including pesticides produced pursuant to an experimental use permit and pesticides, devices, and pesticide active ingredients produced for export.

To maintain records showing product name, EPA Registration Number, Experimental Permit Number if the pesticide is produced under an Experimental Use Permit, amounts per batch and batch identification of all pesticides produced. To also maintain records of production, brand names, receipt, shipment, inventories, advertising, guarantees, disposal, tests, research, and such other records as specified in the section cited.

Retention period: Various.

171.11 Certified commercial pesticides applicators.

To maintain records of specified information relating to the use of restricted use pesticides.

Retention period: 2 years.

172.5 Producers of pesticides produced pursuant to an experimental use permits.

To maintain records in accordance with 40 CFR Part 162.

Guide to Record Retention 1989

180.31 Persons obtaining an experimental permit for use of a pesticide chemical for which a temporary tolerance is established.

To maintain temporary tolerance records of production, distribution, and performance.
Retention period: 2 years.

205.172 Manufacturers of new motorcycle exhaust systems subject to noise emission standards.

To maintain general and individual records as specified in section.
Retention period: 3 years.

224.1 Persons holding permits to allow dumping of material into the ocean waters.

To maintain complete records of materials dumped, time and locations of dumping, and such other records as required in section cited.
Retention period: Not specified.

233.4 Persons submitting applications for a permit to discharge dredged or fill material into State regulated waters.

To keep records of all data used to complete permit application and any other supplemental information.
Retention period: 3 years.

233.7 Persons holding 404 permits to operate and maintain all facilities and systems of treatment and control (and related appurtenances) pursuant to the Dredge or Fill Program under the Clean Water Act.

To maintain all monitoring information, including all calibration and maintenance records and all original strip chart recordings for continuous monitoring instruments, copies of all reports required by the permit, and records of all data used to complete the application for the permit.
Retention period: 3 years.

240.211 Owners and operators of thermal processing facilities and land disposal sites.

To maintain records and monitoring data as required by regulations.
Retention period: Not specified.

240.211-1 Owners and operators of thermal processing facilities and land disposal sites. See 240.211.

240.211-3 Owners and operators of thermal processing facilities and land disposal sites.

To keep operating records in a daily log.
Retention period: Not specified.

241.212-3 Owners/operators of land disposal sites for solid wastes.

To maintain records on major operational problems, complaints or difficulties; qualitative and quantitative evaluation of the environmental impact of the land disposal site with regard to the effectiveness of gas and leachate control; vector control efforts and other data as required by regulation.
Retention period: Not specified.

261.4 Generators of waste samples and owners of operators of laboratories or testing facilities conducting treatability studies. [Added]

(a) To maintain copies of shipping documents; a copy of the contract with the facility conducting the treatability study; documentation showing the amount of waste shipped under the exemption; the name, address, and EPA identification number of the laboratory or testing facility that received the waste; the date the shipment was made; and whether or not unused samples and residues were returned to the generator or sample collector, or if sent to a designated facility, the name of the facility and the EPA identification number, and other information as specified in section cited.
Retention period: 3 years.

(b) To keep, on site, a copy of the treatability study contract and all shipping papers associated with the transport of treatability study samples to and from the facility.
Retention period: 3 years.

262.40 Hazardous waste generators.

(a) To keep a copy of each manifest signed.

Environmental Protection Agency

Retention period: 3 years or until receipt of a signed manifest from facility receiving the hazardous waste.

(b) To keep a copy of each Biennial Report and Exception Report.
Retention period: 3 years from due date of report.

(c) To keep records of any test results, waste analyses, or other determinations that the waste is hazardous.
Retention period: 3 years from the date that the waste was last sent to on-site or off-site treatment, storage, or disposal.

262.57 Primary exporters of hazardous waste.

To maintain copies of each notification of intent to export; each EPA Acknowledgement of Consent; each confirmation of delivery of the hazardous waste from consignee; and each annual report.
Retention period: 3 years.

263.22 Hazardous waste transporter.

To keep a copy of the manifest signed by the generator, transporter, and the next designated transporter or the owner or operator of a designated facility.
Retention period: 3 years from date initial transporter accepted the hazardous waste.

263.22 Hazardous waste transporters (water bulk shipment transporter).

To retain a copy of shipping paper containing all the information for shipments delivered to designated facility by water (bulk shipment).
Retention period: 3 years from date initial transporter accepted the hazardous waste.

263.22 Hazardous waste transporters (initial and final rail transporter).

Initial rail transporter to keep a copy of manifest and shipping paper for shipments by rail within the U.S., final rail transporter to keep a copy of signed manifest (or shipping paper if signed in lieu of manifest by the designated facility) for shipments by rail within the U.S.
Retention period: 3 years from date initial transporter accepted the hazardous waste.

263.22 Hazardous waste transporters (transporter who transmits hazardous waste out of the U.S.

To keep a copy of the manifest indicating when the hazardous waste left the U.S.
Retention period: 3 years from date initial transporter accepted the hazardous waste.

264.15 Owners and operators of all hazardous facilities.

To keep inspection records of the date and time of the inspection, the name of the inspection, a notation of the observations made, and the date and nature of any repairs or other remedial actions.
Retention period: 3 years.

264.16 Hazardous waste transporters.

To maintain personnel training records.
Retention period: For current personnel, until closure of facility; for former personnel, 3 years from date employee left facility.

264.71 Owners and operators of on-site and off-site hazardous waste treatment, storage, and disposal facilities.

To keep a copy of the manifest and shipping paper (if signed in lieu of the manifest at time of delivery) signed by the owner or operator.
Retention period: 3 years from date of delivery.

264.71 Owners and operators of on-site and off-site hazardous waste treatment, storage, and disposal facilities (hazardous waste from a rail or water bulk shipment) transporter.

To keep a copy of each shipping paper and manifest signed by the owner or operator for shipments delivered by rail or water (bulk shipment).
Retention period: 3 years from date of delivery.

264.73 Owners and operators of on-site and off-site hazardous waste treatment, storage, and disposal facilities. [Amended]

To keep a written operating record of the facility.
Retention period: Until at least closure of the facility; monitoring data at

Guide to Record Retention 1989

264.97 disposal facilities: Throughout the post-closure period; records for inspections: 3 years.

264.97 Owners and operators of hazardous waste treatment, storage, and disposal facilities. [Added]

To maintain in the facility operating records, ground-water monitoring data including actual-levels of constituents.

264.98 Owners and operators of hazardous waste treatment, storage, and disposal facilities. [Added]

To maintain records of ground-water analytical data as measured and in a form necessary for the determination of statistical significance under 40 CFR 264.97(h).

264.279 Owners or operators of hazardous waste, storage, and disposal facilities.

To maintain operating records including hazardous waste application dates and rates.
Retention period: 3 years.

264.309 Owners or operators of facilities that dispose of hazardous waste in landfills.

To maintain operating records that include (a) on a map, the exact location and dimensions, including depth, of each cell with respect to permanently survey benchmarks; and (b) the contents of each cell and the approximate location of each hazardous waste type within each cell.
Retention period: 3 years.

264.347 Owners or operators of hazardous waste incinerators.

To maintain inspection log or summary of the date and time of the inspection, the name of the inspector, a notation of the observations made, the date and nature of any repairs or other remedial actions.
Retention period: 3 years.

265.15 Owners or operators of all hazardous waste facilities.

265.16 Hazardous waste transporters.

To maintain personnel training records.
Retention period: For current personnel, until closure of facility; for former personnel, 3 years from date employee left facility.

265.71 Owners and operators of on-site and off-site hazardous waste treatment, storage, and disposal facilities.

To keep a copy of the manifest and shipping paper (if signed in lieu of the manifest at time of delivery) signed by the owner or operator.
Retention period: 3 years from date of delivery.

265.71 Owners and operators of on-site and off-site hazardous waste treatment, storage, and disposal facilities (hazardous waste from a rail or water bulk shipment) transporter.
See 264.71.

265.73 Owners and operators of on-site and off-site hazardous waste treatment, storage, and disposal facilities. [Amended]
See 264.73.

265.94 Owners and operators of on-site and off-site hazardous waste treatment, storage, and disposal facilities.

(a) An owner or operator who does not operate a groundwater quality assessment plan must keep records of analyses of groundwater samples, groundwater surface elevation data, and evaluations of the measurements of groundwater samples.
Retention period: Throughout active life of facility and for disposal facilities, throughout post-closure care period as well.

(b) An owner or operator who operates a groundwater quality assessment plan must keep records of the analyses and evaluations specified in the plan.
Retention period: Throughout the active life of the facility and for disposal facilities, throughout the post-closure care period as well.

Environmental Protection Agency

265.112 Owners and operators of on-site and off-site hazardous waste treatment, storage, and disposal facilities.

To keep a written closure plan at the facility which identifies the steps necessary to completely close the facility at any point during its intended life and at the end of its intended life.
Retention period: Not specified.

265.118 Owners and operators of on- and off-site hazardous waste treatment, storage, and disposal facilities.

To keep a written post-closure plan at the facility which identifies the activities which will be carried on after final closure and the frequency of those activities.
Retention period: Not specified.

265.142 Owners and operators of all hazardous waste facilities.

A facility owner or operator must keep a written estimate of facility closure cost at the facility.
Retention period: Not specified.

265.144 Owners and operators of hazardous waste disposal facilities.

An owner or operator must keep a written estimate of annual cost of post-closure monitoring and maintenance of the facility at the facility.
Retention period: Not specified.

265.279 Owners and operators of hazardous waste land treatment facilities.

To keep hazardous waste application dates and rates in the operating records.
Retention period: 3 years.

265.309 Owners and operators of facilities that dispose of hazardous waste in landfills.

To keep (a) on a map, the exact location and dimensions, including depth, of each cell with respect to permanently surveyed benchmarks; and (b) the contents of each cell and the approximate location of each hazardous waste type within each cell in the operating records.
Retention period: 3 years.

268.6 Owners or operators of hazardous waste treatment, storage, and disposal facilities. [Added]

To maintain and keep on site a copy of the monitoring data collected under the monitoring plan that describes the monitoring program installed at and/or around the unit to verify continued compliance with the conditions of the variance.

268.7 Owners and operators of hazardous waste treatment, storage, and disposal facilities. [Added]

To maintain on-site a copy of the demonstration (if applicable) and certification required for each waste shipment.
Retention period: For at least 5 years from the date that the waste that is the subject of such documentation was last sent to on-site or off-site disposal. The 5-year retention requirement is automatically extended during the course of any unresolved enforcement action regarding the regulated activity or as requested by the Administrator.

270.10 Applicants applying for the hazardous waste permit.

To keep records of all data used to complete permit applications and any supplemental information.
Retention period: 3 years from the date the application is signed.

270.30 Persons holding RCRA permits.

To maintain records of all monitoring information, including all calibration and maintenance records and all original strip chart recordings for continuous monitoring instrumentation, copies of all reports required by permit, and records of all data used to complete the application for the permit.
Retention period: 3 years.

280.20 Owners and operators of new underground storage tank systems. [Added]

To maintain records that demonstrate compliance with design, construction, installation, and notification performance standards.
Retention period: Remaining life of the tank and piping.

-233-

Guide to Record Retention 1989

280.31

280.31 Owners and operators of steel underground storage tank system with corrosion protection. [Added]

For UST systems using cathodic protection, to maintain records of the operation of the cathodic protection to demonstrate compliance with performance standards. These records must provide (a) the results of the last three inspections to ensure the equipment is running properly and (b) the results of testing for proper operation by a qualified cathodic protection tester.

Retention period: For the remaining operating life of the UST.

280.33 Owners and operators of underground storage tank systems. [Added]

To maintain records of each repair.

Retention period: For the remaining operating life of the UST.

280.34 Owners and operators of underground storage tank systems. [Added]

See 280.20, 280.31, 280.33, 280.45, and 280.74.

280.45 Owners and operators of underground storage tank systems. [Added]

(a) To maintain all written performance claims pertaining to any release detection system used, and the manner in which these claims have been justified or tested by the equipment manufacturer or installer.

Retention period: 5 years or for a reasonable period of time determine by the implementing agency, from the date of installation.

(b) To maintain records on the results of any sampling, testing, or monitoring.

Retention period: 1 year, or for a reasonable period of time determined by the implementing agency, except that the results of tank tightness testing must be retained until the next test is conducted.

(c) To maintain written documentation of all calibration, maintenance, and repair of release detection equipment permanently located on-site. Any schedules of required calibration and maintenance provided by the release detection equipment must be retained for 5 years from the date of installation.

280.74 Owners and operators of underground storage tank systems. [Added]

To maintain records on the results of the site investigation conducted at permanent closure.

Retention period: 3 years after completion of permanent closure or change-in-service in one of the following ways: (a) By the owners and operators who took the UST system out of service; (b) by the current owners and operators of the UST system site; or (c) by mailing these records to the implementing agency if they cannot be maintained at the closed facility.

280.107 Owners and operators of underground storage tanks containing petroleum. [Added]

To maintain (a) evidence of all financial assurance mechanisms used to demonstrate financial responsibility; (b) a copy of the instrument worded as specified; (c) when using a financial test or guarantee to maintain a copy of the chief financial officer's letter based on year-end financial statements fo the most recent completed financial reporting year and other such records as specified in cited section.

Retention period: Not specified.

281.32 Owners and operators of approved underground storage tank systems. [Added]

To maintain records of monitoring, testing, repairs, and closure that are sufficient to demonstrate recent facility compliance status.

Retention period: Records demonstrating compliance with repair and upgrading requirements must be maintained for the remaining operating life of the facility, these records must be readily available when requested by the implementing agency.

281.40 Owners and operators of State approved underground storage tank systems. [Added]

To maintain records on operation.

Retention period: Unspecified.

Environmental Protection Agency

372.10 Persons subject to the toxic chemical release reporting; community-right-to-know requirements. [Added]

(a) To maintain a copy of each report submitted; and supporting documentation and materials.

Retention period: 3 years from date of the submission of the report under 40 CFR 372.30.

(b) To maintain all supporting materials and documentation used to determine whether a notice is required and used to developed each required notice under 40 CFR 372.45 and a copy of each notice.

Retention period: 3 years from the date of the submission of a notification under 40 CFR 372.45. Records must be maintained at the facility to which the report applies or from which a notification was provided. Such records must be readily available for purposes of inspection by EPA.

403.12 Publicly owned treatment works (POTWs) and industrial users subject to pretreatment requirements.

To maintain records of monitoring activities and results.

Retention period: 3 years.

600.005-81 Manufacturers of new motor vehicles subject to fuel economy regulations.

To maintain general and individual records related to the sections cited.

Retention period: 5 years after the end of the model year to which the records relate.

600.105-78 Manufacturers of new motor vehicles subject to fuel economy regulations.

See 600.005-81.

600.205-77 Manufacturers of new motor vehicles subject to fuel economy regulations.

See 600.005-81.

600.305-77 Manufacturers of new motor vehicles subject to fuel economy regulations.

See 600.005-81.

600.505-78 Manufacturers of new motor vehicles subject to fuel economy regulations.

See 600.005-81.

704.11 Manufacturers, importers, and processors of chemical substances and mixturers. [Added]

To retain (a) a copy of each report submitted by the person in response to the requirements of section 8 (a) of the Toxic Substances Control Act (TSCA); (b) materials and documentation sufficient to verify or reconstruct the values submitted in the report; (c) a copy of each notice sent by the person, return receipt requested, to that person's customers for the purpose of notifying their customers of the customer's reporting obligations; and (d) all return receipts signed by the person's customer's who received the notice.

Retention period: 3 years.

704.33 Persons who manufacture, import or process P-tert-butylbenzoic acid (P-TBBA), p-tert-butyltoluene (P-TBT) and p-tert-butylbenzaldehyde (P-TBB).

To maintain documentation of information contained in report on data under section 8(a), including information on chemical identity and structure, production, use, exposure, disposal, and health and environmental effects.

Retention period: 5 years from the date of submission of the report.

704.95 Manufacturers and importers of the chemical substances phosphoric acid, (1, 2, ethanediylbis (nitrilo-bis (methylene))) tetrakis-(EDTMPA) and its salts. [Added]

To retain documentation of information contained in reports.

Retention period: 5 years from the date of the submission of the report.

704.105 Persons who propose to manufacture, import,or process hexafluoropropylene oxide (HFPO) for use as an intermediate in the manufacture of fluorinated substances in an enclosed process after Dec. 10, 1987.

To retain documentation of information in the Preliminary Assessment Information Manufacturer's Report.

Retention period: 5 years from date of submission of the reports.

701.142 Persons who manufacture, import, process or propose to manufacture, import, or process HEX-BCH for use as an intermediate in the production of isodrin or endrin, on or after January 2, 1986.

(a) To retain documentation of information contained in the reports.

Retention period: 3 years from the date of the report.

(b) To retain the certification of review to ensure that there has been no reportable event.

Retention period: 3 years from the date of the certification.

710.37 Manufacturers and importers of certain substances included in the Toxic Substances Control Act (TSCA) Inventory.

To maintain records that document any information reported to EPA. For substances that are manufactured or imported at less than 10,000 pounds annually, volume records must be maintained as evidence to support a decision not to submit a report.

Retention period: 4 years beginning with effective date of that reporting period.

717.15 Firms manufacturing or processing chemical substances and mixtures.

To maintain records of significant adverse reactions to human health or the environment alleged to have been caused by chemical substances or mixtures.

Retention period: 30 years for employee health related allegations, and 5 years for all other allegations.

720.78 Manufacturers and importers of new chemical substances subject to the provisions of the Toxic Substances Control Act.

(a) To maintain documentation of information reviewed and evaluated to determine the need to make any notification of risk.

Retention period: 5 years.

(b) To maintain documentation of the nature and method of notification concerning the health and environmental effect of a substance including copies of any labels or written notices used.

Retention period: 5 years.

(c) To maintain documentation of prudent laboratory practices used instead of notification and evaluation.

Retention period: 5 years.

(d) To maintain the names and addresses of any persons other than the manufacturer or importer to whom the substance is distributed, the identity of the substance to the extent known, the amount distributed and copies of notification required under section 720.36(c)(2).

Retention period: 5 years.

(e) Persons manufacturing or importing substance in quantities greater than 100 kilograms per year must maintain records of the identity of the substance to the extent known, the production volume of the substance, and the disposition of the substance.

Retention period: 5 years.

721.40 Manufacturers, importers, or processors submitting a significant new use notice to EPA. [Added]

To maintain documentation of information contained in the significant new use notice.

Retention period: 5 years from the date of the submission of the significant new use notice.

721.47 Persons manufacturing, importing, or processing chemical substances for significant uses in small quantities solely for research and development. [Added]

To retain (a) copies of citations to information reviewed and evaluated to determine the need to make any notification of risk; (b) documentation of the nature and method of notification including copies of any labels or written notices used; (c) documentation of prudent laboratory practices used instead of notification and evaluation; and (d) the names and addresses of any persons other than the manufacture, importer, or processor, to whom the substance is distributed, the amount distributed, and copies of the notification required.

Retention period: 5 years.

721.100 Manufacturers, importers, and processors of disubstituted diamino anisole. [Redesignated from 721.120]

In addition to the requirements of section 721.40, to maintain records which include the results of any determination that gloves are impervious, the names of persons required to wear gloves, and copies of labels.

Retention period: 5 years from the date of creation of the record.

721.520 Manufacturers, importers, and processors of substituted polyglycidyl benzeneamine (P-83-394). [Redesignated from 721.180]

In addition to the requirements of section 721.40, to maintain records which include the names of persons informed of the hazards associated with the substance, the names of any transferee and the dates of any transfers of containers which are labeled, and the method used to determine that protective gloves are impervious to the substance and the date and results of the determination.

Retention period: 5 years from the date of creation of the record.

721.800 Manufacturers, importers, and processors of Dicarboxylic acid monoester. [Redesignated from 721.290]

In addition to the requirements of section 721.40, to maintain records including the names of persons required to wear protective clothing, and the name and address of each person to whom the substance is sold or transferred and the date of such sale or transfer.

Retention period: 5 years from the date of creation of the record.

721.1875 Manufacturers, importers, and processors of Substituted methylpyridine (P-83-21, P-83-49, and P-83-272). [Redesignated from 721.615]

In addition to the requirements of section 721.40, to maintain records including the names of persons required to wear protective clothing and/or equipment, records of respirator fit tests for each person required to wear a respirator, and the names and addresses of persons to whom any of these substances are sold or transferred and the date of such sale or transfer.

Retention period: 5 years from the date of creation of the record.

721.2100 Manufacturers, importers, and processors of derivative of tetrachloroethylene (P-62-684). [Redesignated from 721.975]

In addition to the requirements of section 721.40 to maintain records including the names of persons required to wear protective equipment, the names and addresses of any person to whom the substance is sold or transferred and the dates of such sale or transfer, records of respirator fit tests for each person required to wear a respirator, and the method for determining that the gloves required under this section are impervious to the substance, the date(s) of such determination, and the results of that determination.

Retention period: 5 years from the date of the creation of the record.

723.50 Manufacturers of new chemical substances manufactured in quantities of 1,000 kilograms or less per year under low volume exemption.

To maintain records of (a) the annual production volume of the new chemical substance under the exemption and (b) documentation of information in the exemption notices and compliance with the terms of the regulations.

Retention period: 5 years.

723.175 Manufacturers and processors of new chemical substances used in or for the manufacture of processing of instant photographic and peel-apart film articles under instant photographic chemical exemption.

To keep records on annual production volume, exposure monitoring, worker's training and exposure, and method of treatment.

Retention period: 30 years.

723.250 Manufacturers of new polymers under polymer exemption.

To maintain records of production volume for the first 3 years of manufacture, the date of commencement of manufacture, and documentation of this information and any other infor-

761.30

mation provided in the limited premanufacture notice.
Retention period: 5 years.

761.30 Owners or operators of PCB Transformers in use or stored for reuse.

To maintain records of inspections and maintenance history, including leaks, repairs, replacement, cleanup and containment.
Retention period: 3 years after disposal of transformer.

761.30 Owners or operators of railroad transformers using PCBs. [Amended]

(a) To maintain records of concentration of PCBs in dielectric fluid after servicing railroad transformer for purposes of reducing PCB concentration in dielectric fluid.
Retention period: Until January 1, 1991.

(b) To maintain at the facility documentation to support the reason for the emergency installation of a PCB transformers. Documentation must be completed within 30 days after installation of the PCB Transformer.

761.30 Owners or operators of heat transfer systems that ever contained PCBs in heat fluid in concentrations greater than 50 ppm.

To maintain records of data obtained from required sampling of PCB concentration in heat transfer fluid.
Retention period: 5 years after PCB concentration in fluid reaches 50 ppm.

761.30 Owners or operators of hydraulic systems that ever contained PCBs at concentrations above 50ppm.

To maintain records of data obtained from required test sampling of PCB concentration in hydraulic fluid.
Retention period: 5 years after PCB concentration in fluid reaches 50 ppm.

761.60 Owners or operators of high efficiency boilers used to dispose of mineral oil dielectric fluid and other liquids containing between 50 and 500 ppm PCBs.

To maintain records of quantities of mineral oil dielectric fluid burned each month, and data from monitoring of combustion.
Retention period: 5 years.

Guide to Record Retention 1989

761.75 Owners or operators of facilities used to dispose of PCBs.

To maintain records of all disposal operations.
Retention period: 5 years.

761.80 Processors and distributors of PCBs and PCB items.

To maintain records of activities.
Retention period: 5 years.

761.80 Processors and distributors of PCB in small quantities for research and development having class exemption.

To maintain records of PCB activities.
Retention period: 5 years.

761.125 Owners of spilled PCB's.

(a) At the completion of cleanup, to document the cleanup with records and certification of decontamination.
Retention period: 5 years.

(b) If cleanup is delayed because of adverse weather conditions, lack of access due to physical impossibility or emergency operating conditions, to maintain records documenting the fact that circumstances precluded rapid response.
Retention period: Not specified.

761.180 Owners or operators of facilities used to dispose of PCBs.

To maintain records on disposal, storage, chemical waste landfills, incineration, high efficiency broilers, and other documentation as specified in section.
Retention period: 5 years; chemical waste landfill data 20 years.

761.193 Persons importing, manufacturing, processing, distributing in commerce or using chemicals containing inadvertently generated PCBs.

To maintain records of actual monitoring of PCB concentrations.
Retention period: 3 years after a process ceases operation or importing ceases, or for 7 years, whichever is shorter.

763.93 Local educational agencies identifying asbestos-containing materials in schools.

To maintain in its administrative office a complete updated copy of the management plan. In addition, to maintain records on response actions, operations and maintenance, and training and periodic surveillance as part of the management plan.
Retention period: 3 years.

763.94 Local educational agencies identifying asbestos-containing materials in schools.

To maintain records on response actions, operations and maintenance, training and periodic surveillance, and management plans.
Retention period: 3 years.

763.114 Local education agencies identifying friable-asbestos-containing materials in schools.

To maintain results of inspections and analyses; copies of Notice to School Employees; blueprint, diagram or written description of the buildings and other information as specified in the section.
Retention period: Not specified.

763.121 Employers of employees covered by the EPA asbestos abatement worker protection rule.

(a) To maintain records of objective data for exempted data when relying on objective data that demonstrate that products made from or containing asbestos fibers of asbestos in concentrations at or above the action level under the expected conditions of processing, use, or handling to exempt such operations from the initial monitoring requirements.
Retention period: Duration of the employer's reliance upon such objective data.

(b) To maintain records of all measurements taken to monitor employee exposure to asbestos.
Retention period: At least 30 years.

(c) To maintain employee medical surveillance records.
Retention period: Duration of employee employment plus 30 years.

(d) To maintain all employee training records.
Retention period: 1 year beyond the last date of employment.

Environmental Protection Agency

763.121 Appendix C Employers of employees covered by the EPA asbestos abatement worker protection rule.

To maintain summary of all qualitative fit test results.
Retention period: 3 years.

Part 763, Appendix A to Subpart E Laboratories which analyze asbestos bulk samples and asbestos air samples.

To maintain log of all pertinent sampling information and appropriate logs or records verifying compliance with the mandatory quality insurance procedures.
Retention period: Not specified.

Part 763, Appendix D to Subpart E Transporters of asbestos waste.

To maintain as evidence of receipt at the disposal site a copy of the chain of custody form signed by the disposal site operator.
Retention period: Not specified.

792.29 Toxic substances control testing facilities.

To maintain a current summary of training and experience and job description for each individual engaged in or supervising the conduct of a study.
Retention period: 10 years.

792.31 Toxic substances control testing facility management.

To document and maintain such action as raw data, records on the replacement of the study director if it becomes necessary to do so during the conduct of a study.
Retention period: 10 years.

792.33 Toxic substances control testing facilities study directors.

To maintain and verify (a) all experimental data, including observations of unanticipated responses of test systems; (b) notes on unforeseen circumstances that may affect the quality and integrity of the study when they occur; and (c) documentation of the corrective action taken.
Retention period: In accordance with 40 CFR 792.195.

792.35 Toxic substances control testing facility quality assurance units.

To maintain a copy of a master schedule sheet of all studies conducted at the testing facility; copies of all protocols pertaining to all studies for which the unit is responsible; and copies of written and properly signed records of each periodic inspection.

Retention period: Indefinitely.

792.63 Toxic substances control testing facility quality assurance units.

To maintain records of all inspection, maintenance, testing, calibrating, standardizing operations, and nonroutine repairs performed on equipment as a result of failure and malfunction.

Retention period: 10 years.

792.81 Toxic substances control testing facilities.

To maintain a historical file of standard operating procedures and all revisions thereof, including the dates of such revisions.

Retention period: In accordance with 40 CFR 792.195.

792.90 Toxic substances control testing facilities.

To maintain as raw data documentation of the analyses of feed and water used for the animals to ensure that contaminants known to be capable of interfering with the study and reasonably expected to be present in such feed or water are not present at levels above those specified in protocol and to maintain documentation of any use of pest control materials.

Retention period: In accordance with 40 CFR 792.195.

792.105. Toxic substances control testing facilities.

(a) For each batch, to maintain documentation on the indentity, strength, purity, and composition or other characteristics which will appropriately define the test or control substance.

(b) To maintain documentation on the methods, fabrication, or derivation of the test and control substances.

(c) For studies of more than 4 weeks' duration, to reserve samples from each batch of test-control substances.

Retention period: In accordance with 40 CFR 792.195.

792.185 Sponsors and toxic substances control testing facilities.

For each study, to maintain a copy of the final report and of any amendments to it.

Retention period: In accordance with 40 CFR 792.195.

799.10 Persons who manufacture or intend to manufacture (including import and/or persons who process or intend to process a chemical substance or mixture (DETA) under the specific chemical test rule.

To maintain all raw data, documentation, records, protocols, specimens and reports generated as a result of study in accordance with the TSCA Good Laboratory Practice Standards (GLP's) in 40 CFR Part 792.

Retention period: In accordance with 40 CFR 792.195.

January 1, 1990

Supplement to
Guide to Record
Retention Requirements

CFR Title 40

ENVIRONMENTAL PROTECTION AGENCY

40 CFR

35.6250 Recipients of CERCLA-funded cooperative agreements and Superfund State Contracts. [Added]

(a) To maintain a recordkeeping system that consists of complete site-specific files containing documentation of costs incurred.

(b) To maintain records to comply with the requirements of 40 CFR 35.6700, 35.6705, and 35.6710 and requirements of source documentation described in 40 CFR 31.20(b)(6).

Retention period: 10 years following submission of the final Financial Status Report for the site, or until resolution of all issues arising from litigation, claim, negotiation, audit, cost recovery, or other actions, whichever is later. Written approval must be obtained from the EPA award official before destroying any records.

35.6700 Recipients of CERCLA-funded cooperative agreements and Superfund State Contracts. [Added]

(a) To maintain project records by site and activity.

(b) To maintain property, financial and procurement records.

(c) To maintain time and attendance records and supporting documentation; documentation of compliance with statutes and regulations that apply to the project; and the number of site-specific technical hours spent to complete each pre-remedial product.

Retention period: 10 years following submission of the final Financial Status Report for the site, or until resolution of all issues arising from litigation, claim, negotiation, audit, cost recovery or other actions, whichever is later. Written approval must be obtained from the EPA award official before destroying any records.

35.6705 Recipients of CERCLA-funded cooperative agreements and Superfund State Contracts. [Added]

To maintain all financial and programmatic records, supporting documents, statistical records, and other records which are required by 40 CFR 35.6700, program regulations, or the cooperative agreement, or are otherwise reasonably considered as pertinent to program regulations or the cooperative agreement.

Retention period: 10 years following submission of the final Financial Status Report for the site, or until resolution of all issues arising from litigation, claim, negotiation, audit, cost recovery, or other actions, whichever is later. Written approval must be obtained from EPA award official before destroying any records.

60.49b Owners or operators of industrial-commercial-institutional steam generating facilities. [Amended]

(a) If monitoring of steam generating unit operating condition plan is approved, to maintain records of predicted nitrogen oxide emission rates and the monitored operating conditions, including steam generating unit load, identified in the plan

(b) To maintain records of the amounts of all fuels fired during each day and calculate the annual capacity factor individually for coal, distillate oil, residual oil, natural gas, wood, and municipal-type solid waste for each calendar year.

(c) To maintain records of the nitrogen content of the oil residual combusted in the affected facility and calculate the average fuel nitrogen content on a per calendar quarter basis.

(d) To maintain records of opacity for facilities subject to the opacity standard under 40 CFR 60.43b.

(e) To maintain records on the calendar date, the average hourly nitrogen oxides emission rates measured or predicted and other information as specified in section cited for each steam generating unit operating day for facilities subject to nitrogen oxide standards under 40 CFR 60.44(b).

Retention period: 2 years following date of record.

(f) To maintain records of the following information for each steam generating unit operating day: (1) Calendar date; (2) the number of hours of operation; and (3) a record of the hourly steam load.

60.545 Owners or operators of tread end cementing operation and green tire-spraying operation using water-based cements or sprays containing less than 1.0 percent by weight of VOC. [Added]

To maintain records of formulation data or the results of Method 24 analysis conducted to verify the VOC contents of the spray.

Retention period: Not specified.

60.744 Owners or operators of new, modified and reconstructed facilities that perform polymeric coating of supporting substrates. [Added]

To maintain records of the measurements and calculations required in 40 CFR 60.743 and 60.744.

Retention period: For at least 2 years following the date of the measurements anbd calculations.

60.747 Owners or operators of new, modified and reconstructed facilities that perform polymeric coating of supporting substrates. [Added]

To maintain records documenting compliance by the methods described in 40 CFR 60.743(a)(1), (a)(2), (a)(4), (b), or (c).

Retention period: At least 2 years.

61.25 Owners or operators of underground uranium mines. [Added]

To maintain records documenting the source of input parameters including the results of all measurements upon which they are based, the calculations and/or analytical methods used to derive values for input parameters, and the procedure used to determine compliance. In addition, the documentation should be sufficient to allow an independent auditor to verify the accuracy of the determination made concerning the facility's compliance with the standard.

Retention period: 5 years and records must be made available for inspection by the Administrator or his authorized representative.

-238-

61.26 Owners or operators of underground uranium mines. [Revised; record retention requirements now in 61.25]

61.123 Owners and operators of calciners and nodulizing kilns at elemental phosphorous plants. [Revised; record retention requirements now in 61.124]

61.124 Owners or operators of calciners and nodulizing kilns at elemental phosphorus plants. [Added]

See 40 CFR 61.25.

61.133 Owners or operators of coke by-product recovery plants. [Added]

(a) To maintain records pertaining to the design of control equipment installed to comply with 40 CFR 61.132 through 6.134.

(b) To maintain records pertaining to sources subject to 40 CFR 61.132 and 61.133. Such records shall contain the date of the inspection and the name of the inspector; a brief description of each visible defect in the source or control equipment and the method and date of repair of the defect; the date of attempted and actual repair and method of repair of the leak; and a brief description of any system abnormalities found during the annual maintenance inspection, the annual maintenance inspection, the repairs made, the date of attempted repair, and the date of actual repair.

Retention period: 2 years following each semiannual (and other) inspection and each annual maintenance inspection.

61.204 Owners and operators of the phosphogypsum that is produced as a result of phosphorus fertilizer production and all that is contained in existing phosphogypsum stacks. [Added]

See 40 CFR 61.25.

61.224 Owners and operators of all sites that are used for the disposal of uranium mill tailings. [Added]

See 40 CFR 61.25.

61.246 Owners or operators of sources intended to operate in volatile hazardous air pollutant (VHAP) service. [Added]

(a) To keep records in a log of each leak as specified in 40 CFR 61.242-2, 61.242-3, and 61.242-7.

Retention period: 2 years.

(b) To maintain records, in a log, pertaining to all equipment subject to the requirements in 40 CFR 61.242-1 to 40 CFR 61.242-11 and all other records as specified in cited section.

61.255 Owners or operators of facilities byproduct materials during and following the processing of uranium ores, commonly referred to as uranium mills and their associated tailings. [Added]

See 40 CFR 61.25

61.276 Owners or operators with a storage vessel subject to the national emission standard for benezene emissions. [Added]

(a) To keep readily accessible records showing the dimensions of the storage vessel and an analysis showing the capacity of the storage vessel.

Retention period: As long as the storage vessel is in operation.

(b) To keep records pertaining to closed vent system and control devices in a readily accessible location.

Retention period: 2 years.

80.27 Distributors, resellers, carriers, retailers, and wholesale purchaser-consumers of gasoline and alcohol blends volatility. [Added]

To maintain each invoice, loading ticket, bill lading, delivery ticket and other documents which accompany the shipment of such gasoline. Such documents shall be available for inspection by the Administrator or authorized representative during such period.

Retention period: 1 year.

82.13 Importers of certain chlorofluorocarbons (CFC's) and brominated compounds (halons) to reduce the risk of stratospheric ozone depletion. [Amended]

To maintain records on (a) the quantity of each controlled substances imported, either alone or in mixtures; (b) the quantity of each controlled substances were imported; (c) the port of entry through which the controlled substances passed; (d) the country from which the imported controlled substances were imported; the port of exit; and other information as specified in cited section.

Retention period: 3 years.

86.090-14 Small-volume manufacturers of light-duty vehicles, light-duty trucks, and heavy-duty engines subject to air pollution controls. [Added]

To maintain records of all the information required by 40 CFR 86-090-21.

86.090-24 Manufacturers of new motor vehicles and new motor vehicle engines subject to air pollution controls. [Added]

To maintain and make available to EPA Administrator upon request, the engineering evaluation, including any test data used to support the deletion of optional equipment from the test vehicles.

86.090-26 Manufacturers of light-duty vehicles subject to air pollution controls. [Added]

(a) To maintain and provide to EPA Administrator, a record of the rationale used in making for engine family, the mileage at which the engine-steam combination is stabilized for emission-data testing determination.

(b) To retain records of all information concerning all emission tests and maintenance, including vehicle alterations to represent other vehicle selections whenever a manufacturer intends to operate and test a vehicle which may be used for emission or durability data.

86.107-90 Manufacturers of petroleum-fueled and methanol-fuel light-duty vehicles and light-duty trucks. [Added]

(a) To maintain permanent records of results at the initiation and termination of each diurnal or hot soak in measuring hydrocarbon (hydrocarbons plus methanol as appropriate).

(b) For the methanol sample to maintain permanent records of the following: (1) The volumes of deionized water introduced into each impinger; (2) the rate and time of sample collection; (3) the volumes of each sample introduced into the gas chromatograph; (4) the flow rate of carrier gas through the carrier; and (5) the chromatogram of the analyzed sample.

86.142-90 Manufacturers of 1977 and later motor year new light-duty vehicles and new light-duty trucks subject to emission test procedures. [Added]

To maintain for each test: (a) Test number; (b) system or device tested (brief description); (c) date and time of day for each part of the test schedules; (d) instrument operated; (e) driver or operator; (f) vehicle ID number, manufacturer, model year, standard, engine family, evaporative emissions family, basic engine description; and such other information as cited in section.

86.605 Manufacturers of new gasoline-fueled and diesel light-duty vehicles and new gasoline-fueled and diesel light-duty trucks subject to selective enforcement auditing procedures required by air pollution control regulations. [Amended]

To maintain general and individual records relating to vehicle emission tests performed pursuant to test orders as specified in the section cited.

Retention period: 1 year after completion of tests.

86.609-88 Manufacturers of new gasoline-fueled and diesel light-duty vehicles and new gasoline-fueled and dieseled light-duty trucks subject to selective enforcement auditing procedures required by air pollution control regulations. [Added]

To maintain equivalency documentation if using an equivalent method when measuring the temperature of the test fuel at other than the approximate mid-volume of the fuel tank and when draining the test fuel from other than the lowest point of the tank.

Retention period: 1 year after completion of all testing in response to a test order.

86.1005-84 Manufacturers of new gasoline-fueled and diesel heavy-duty vehicles and new gasoline-fueled and diesel heavy-duty trucks subject to selective enforcement auditing procedures required by air pollution control regulations. [Redesignated as 86.1005-88 and amended]

To maintain general and individual records relating to vehicle emission tests performed pursuant to test orders as specified in the section cited.

Retention period: 1 year after completion of tests.

86.1005-90 Manufacturers of new petroleum-fueled or methanol-fueled heavy duty or engine or light-duty trucks. [Added]

To maintain testing and auditing records as specified in section cited.

Retention period: 1 year after completion of all testing in response to a test order.

86.1008-88 Manufacturers of new gasoline-fueled and diesel heavy-duty vehicles and new gasoline-fueled and diesel heavy-duty trucks subject to selective enforcement auditing procedures required by air pollution control regulations. [Added]

See 40 CFR 86.608-88.

86.1008-90 Manufacturers of new petroleum-fueled or methanol-fueled heavy-duty engines or light-duty trucks. [Revised]

To maintain and make available to the EPA Administrator upon request, equivalency test documentation.

86.1242-90 Manufacturers of new gasoline-fueled and methanol-fueled heavy duty vehicles. [Added]

See 40 CFR 86.142-90.

122.21 Persons holding or applying for permits to discharge wastes pursuant to the national pollutant discharge elimination program. [Amended]

To maintain records of all information resulting from monitoring activities and relating to all sludge-related application data and other such information as indicated in section cited.

Retention period: 5 years (or longer as required by 40 CFR Part 403), except for records of monitoring information— 3 years.

142.14 State agencies having primary enforcement responsibilities over public water. [Revised]

To maintain records of tests, measurement, analyses, decisions, and determinations performed on each public water system to determine compliance with applicable provisions of State primary drinking water regulations.

Retention period: (a) Records of turbidity measurements—for less than 1 year; (b) records of disinfectant residual measurements and other parameters necessary to document disinfection effectiveness and applicable reporting requirements—not less than 1 year; (c) records of decisions—40 years or until 1 year after the decision is reversed or revised; (d) records of any determination that a public water system supplied by a surface water source or a ground water source under the direct influence of surface water is not required to provide filtration treatment—40 years or until withdrawn; (e) records of analyses for contaminants other than microbiological contaminants (including total coliform, fecal coliform, and hetertrophic plate concentration, other parameters necessary to determine disinfection effectiveness (including temperature and pH measurements) and turbidity— 40 years; (f) records of microbiological analyses of repeat or special samples—1 year in the form of actual laboratory reports or in an appropriate summary form; and (g) records of decisions made pursuant to the total coliform provisions of 40 CFR Part 141—5 years.

160.29 Testing facilities conducting studies that support applications for research or marketing permits for pesticides regulated by EPA. [Revised]

To maintain a current summary of training and experience and job description for each individual engaged in or supervising the conduct of a study.

Retention period: 5 years.

160.63 Testing facilities conducting studies that support applications for research or marketing permits for pesticides regulated by EPA. [Revised]

To maintain written records of all inspection, maintenance, testing, calibrating, and/or standardizing operations. Also to maintain written records of nonroutine repairs performed on equipment as a result of failure and malfunction.

Retention period: 2 years.

160.81 Testing facilities conducting studies that support applications for research or marketing permits for pesticides regulated by EPA. [Added]

To maintain historical file of standard operating procedures and all revisions thereof, including the dates of such revisions.

Retention period: In accordance with 40 CFR 160.195.

160.120 Testing facilities conducting studies that support applications for a research or marketing permits for pesticides regulated by EPA. [Added]

To maintain with the protocol records of all changes in or revisions of an approved protocol and the reasons therefore.

Retention period: In accordance with 40 CFR 160.195.

160.195 Testing facilities conducting studies that support applications for research or marketing permits for pesticides regulated by EPA. [Added]

(a) To maintain documentation records, raw data, and specimens pertaining to a study and required to be retained.

Retention period: (1) In the case of a study used to support an application for a research or marketing permit approved by EPA, the period during which the sponsor holds any research or marketing permit to which the study is pertinent.(2) A period of at least 5 years following the date on which the results of the study are submitted to the EPA in support of an application for research marketing year. (3) In other situations (e.g. where the study does not result in the submission of the study in support of an application for a research or marketing permit), a period of at least 2 years following the date on which the study is completed, terminated, or discontinued.

(b) Wet specimens, samples of test, control, or reference substances, and specially prepared material which are relatively fragile and differ markedly in stability land quality during storage shall be retained only as long as the quality of the preparation affords evaluation.

(c) To maintain the master schedule sheet, copies of protocols and records of quality assurance inspections in accordance with 40 CFR 160.195(b).

(d) To maintain summaries of training and experience and job description in accordance with 40 CFR 160.195(b).

259.54 Generators of medical waste, including generators of less than 50 pounds per months. [Added]

(a) To keep a copy of each tracking form signed in accordance with 40 CFR 259.52

Retention period: For at least 3 years from the date the waste was accepted by the initial transporter.

(b) To retain a copy of all exception reports required to be submitted under 40 CFR 259.55.

(c) To maintain a shipment log at the original generation point.

Retention period: For a period of 3 years from the date the waste was shipped.

(d) To maintain a shipment log at each central collection point and other such records as specified in cited section.

Retention period: For a period of 3 years from the date that regulated medical waste was accepted from each original generation point.

259.76 Transporters of medical waste. [Added]

To retain a copy of each tracking form in accordance with 40 CFR 259.77.

259.77 Transporters of regulated medical waste. [Added]

(a) To keep a copy of the tracking form signed by the generator, himself, the previous transporter (if applicable), and the next party, which may be one of the following: Another transporter or the owner or operator of an intermediate handling facility, or destination facility.

Retention period: For a period of 3 years from the date the waste was accepted by the next party.

(b) For regulated medical waste that is not accompanied by a generator-initiated tracking form, to retain a copy of all transporter-initiated tracking forms and consolidation logs.

Retention period: For a period of 3 years from the date the waste was accepted by the transporter.

(c) For any regulated medical waste that was received by the transporter accompanied by a tracking form and consolidated by a tracking form and consolidated or remanifested by the transporter to another tracking form, to retain (1) a copy of the generator-initiated tracking form signed by the transporter; (2) a copy of the transporter-initiated tracking form signed by the intermediate handler or destination facility.

Retention period: (1) 3 years from the date the waste was accepted by the transporter; and (2) 3 years from the date the waste was accepted by the intermediate handler or destination facility.

(d) To retain a copy of each transporter report required by 40 CFR 259.78.

Retention period: 3 years after the date of submission.

259.81 Owners or operators of facilities including destination and intermediate facilities receiving regulated medical waste generated in a Covered State. [Added]

(a) To retain a copy of each tracking form in accordance with 40 CFR 259.83.

(b) To retain a copy of the tracking form or shipping papers if signed in lieu of the tracking form.

Retention period: For at least 3 years from the date of acceptance of the regulated medical waste.

259.83 Owners or operators of destination facilities or intermediate handlers receiving regulated medical waste generated in a Covered State. [Added]

(a) To maintain (1) copies of all tracking forms and logs; (2) the name and State permit or identification number of each generator who delivered waste to the destination facility or intermediate handler, if the State does not issue permit or identification numbers then the generator's address; and (3) copies of all discrepancy reports.

(b) To maintain the following information for each shipment of regulated medical waste accepted: (1) The date the waste was accepted; (2) the name and State permit or identification number of the generator who originated shipment. If the State does not issue permit or identification numbers, then the generator's address; (3) the total weight of the regulated medical waste accepted from the originating generator; and (4) the signature of the individual accepting the waste.

Retention period: 3 years from the date the waste was accepted.

259.90 Persons engaged in rail transportation of regulated medical waste generated in a Covered State. [Added]

To retain a copy of the tracking forms and rail shipping papers in accordance with 40 CFR 259.77.

501.15 Applicants for the State sludge management program. [Added]

To maintain records of all data used to complete permit applications and any supplemental information.

Retention period: 5 years from the date the application is signed or as required by 40 CFR Part 503.

501.15 Persons holding permits under the State Sludge Management Program. [Added]

To retain records of all monitoring information, copies of all reports required by the permit, and records of all data used to complete the application for the permit.

Retention period: At least 5 years from the date of the sample, measurement, report or application, or longer as required by 40 CFR Part 503.

This period may be extended by request of the Director at any time.

721.17 Manufacturers, importers, and processors of chemical substances which EPA has determined are significant new uses under certain provisions of the Toxic Substances Control Act. [Added]

To maintain documentation of information contained in that person's significant new use notice.

Retention period: For a period of 5 years from the date of the submission of the significant new use notice.

721.72 Manufacturers, importers, and processors of substances; significant new use rules. [Added]

To maintain records documenting establishment and implementation of a hazard communication program. The hazard communication program will, at a minimum, describe how the requirements of this section for labels, MSDs, and other forms of warning material will be satisfied.

Retention period: 5 years from the date of creation.

721.125 Manufacturers, importers, and processors of substances; significant new use rules. [Added]

(a) To maintain records documenting the manufacture and importation volume of the substance and the corresponding dates of manufacture and import.

(b) To maintain records documenting volumes of the substance purchased in the United States by processors of the substance, names, and addresses of suppliers, and corresponding dates of purchases.

(c) To maintain records documenting establishing and implementation of a program for the use of any applicable personal protective equipment required under 40 CFR 721.63.

Retention period: 5 years from the date of their creation.

721.557 Manufacturers, importers, and processors of mixture of 1,3-benzenediamine, 2-methyl-4,6-bis (methylthio)- and 1,3- benzenediamine, 4-methyl-2,6-bis (methylthio); significant new uses; Toxic Substances Control Act. [Added]

In addition to the requirements of 40 CFR 721.17, to maintain the following records: (a) Any determination that gloves are impervious to the substance; (b) names of persons who have attended safety meetings; the dates of such meetings, and copies of any written information provided; copies of any MSDs used, names and addresses of all persons to whom the PMN substance is sold or transferred including shipment destination address if different, the date of each sale or transfer, and the quantity of substance sold or transferred on such date; copies of any labels used; and other information as specified in cited section.

Retention period: 5 years after the date the records are created.

761.180 Owners or operators of facilities used to dispose of PCBs. [Amended]

To maintain annual records on disposal, storage, chemical waste landfills, incineration, high efficiency broilers, and other documentation as specified in section.

Retention period: 3 years; chemical waste landfill data 20 years.

761.209 Transporters of PCB waste. [Added]

To maintain a copy of the manifest signed by the generator, transporter, and the next designated transporter, if applicable, or the owner or operator of the designated commercial storage or disposal facility.

Retention period: 3 years from the date the PCB waste was accepted by the initial transporter. Record retention period may be extended automatically during the course of any outstanding enforcement action regarding the regulated activity.

761.209 Generators of PCB waste. [Added]

To keep a copy of each manifest signed until a signed copy from the designated commercial storage or disposal facility which received the PCB waste is received.

Retention period: 3 years from the date the PCB waste was accepted by the initial transporter. Record retention period may be extended automatically during the course of any outstanding enforcement action regarding the regulated activity.

761.209 Water (bulk shipment) transporters of PCB waste. [Added]

To retain a copy of the shipping papers for shipments of PCB waste delivered to designated commercial storage or disposal facility by water (bulk shipment).

Retention period: 3 years from the date the PCB waste was accepted by the initial transporter. Record retention period may be extended automatically during the course of any outstanding enforcement action regarding the regulated activity.

761.209 Initial rail transporters of PCB waste. [Added]

To maintain a copy of the manifest and the shipping paper required to accompany the PCB waste.

Retention period: 3 years from the date the PCB waste was accepted by the initial transporter. Record retention period may be extended during the course of any outstanding enforcement action regarding the regulated activity.

761.209 Final rail transporters of PCB waste. [Added]

To keep a copy of the signed manifest, or the required shipping paper if signed by the designated facility in lieu of the manifest.

Retention period: 3 years from the date the PCB waste was accepted by the initial transporter. Record retention period may be extended automatically during the course of any outstanding enforcement action regarding the regulated activity.

760.218 Disposal facilities, generators and commercial storers of PCB waste. [Added]

To keep a copy of each Certificate of Disposal.

Retention period: See 40 CFR 761.180.

763.178 Persons producing an asbestos-containing product that is subject to a labeling requirements. [Added]

To maintain a copy of the label used in compliance.

Retention period: 3 years after the effective date of the ban on distribution in commerce for the product which the labeling requirements apply.

763.178 Persons producing an asbestos-containing product that is subject to a manufacture, importation and/or processing ban. [Added]

To maintain the results of the inventory for the banned product.

Retention period: 3 years after the effective date of the ban on manufacture, importation, and processing.

763.178 Persons whose asbestos-producing activities are subject to the manufacture, importation, processing and distribution in commerce bans. [Added]

To maintain all commercial transactions regarding the product including the date of purchases and sale and the quantities purchased or sold.

Retention period: 3 years after the effective date of the ban on distribution in commerce for a product.

792.29 Toxic substances control testing facilities. [Revised]

To maintain a current summary of training and experience and job description for each individual engaged in or supervising the conduct of a study.

Retention period: 10 years.

792.31 Toxic substances control testing facility management. [Revised]

To document and maintain such action as raw data, records on the replacement of the study director if it becomes necessary to do so during the conduct of a study.

Retention period: 10 years.

792.33 Toxic substances control testing facilities study directors. [Revised]

To maintain and verify (a) all experimental data, including observations of unanticipated responses of test systems; (b) notes on unforeseen circumstances that may affect the quality and integrity of the study when they occur; and (c) documentation of the corrective action taken.

Retention period: In accordance with 40 CFR 792.195.

792.35 Toxic substances control testing facility quality assurance units. [Revised]

To maintain a copy of a master schedule sheet of all studies conducted at the testing facility; copies of all protocols pertaining to all studies for which the unit is responsible; and copies of written and properly signed records of each periodic inspection.

Retention period: Indefinitely.

792.63 Toxic substances control testing facility quality assurance units. [Revised]

To maintain records of all inspection, maintenance, testing, calibrating, standardizing operations, and nonroutine repairs performed on equipment as a result of failure and malfunction.

Retention period: 10 years.

792.81 Toxic substances control testing facilities. [Revised]

To maintain a historical file of standard operating procedures and all revisions thereof, including the dates of such revisions.

Retention period: In accordance with 40 CFR 792.195.

792.90 Toxic substances control testing facilities. [Revised]

To maintain as raw data documentation of the analyses of feed and water used for the animals to ensure that contaminants known to be capable of interfering with the study and reasonably expected to be present in such feed or water are not present at levels above those specified in protocol and to maintain documentation of any use of pest control materials.

Retention period: In accordance with 40 CFR 792.195.

792.105 Toxic substances control testing facilities. [Revised]

(a) For each batch, to maintain documentation on the identity, strength, purity, and composition or other characteristics which will appropriately define the test or control substance.

(b) To maintain documentation on the methods, fabrication, or derivation of the test and control substances.

(c) For studies of more than 4 weeks' duration, to reserve samples from each batch of test-control substances.

Retention period: In accordance with 40 CFR 792.195.

792.185 Sponsors and toxic substances control testing facilities. [Revised]

For each study, to maintain a copy of the final report and of any amendments to it.

Retention period: In accordance with 40 CFR 792.195.

792.190 Toxic substances control testing facilities. [Added]

(a) To maintain all raw data, documentation, records, protocols, specimens, and final reports generated as a result of a study.

(b) To maintain correspondence and other documents relating to interpretation and evaluation of data, other than those documents contained in the final report.

Retention period: See 40 CFR 792.195.

792.195 Toxic substances control testing facilities—retention period. [Added]

(a) Documentation records, raw data, and specimens pertaining to a study and required to be retained by 40 CFR Part 792 shall be retained in the archive(s) for a period of at least 10 years following the effective date of the applicable final test rule.

(b) In the case of negotiated testing agreement, documentation records, raw data, and specimens pertaining to a study and required to be retained by 40 CFR Part 792 shall be retained in the archive(s) for a period of at least 10 years following the publication date of the acceptance of a negotiated test agreement.

(c) In the case of testing submitted under section 5, documentation records, raw data, and specimens pertaining to a study and required to be retained under 40 CFR Part 792 shall be retained in the archive(s) for a period of at least 5 years following the date on which the results of the study are submitted to the agency.

(d) Wet specimens, samples of test, control or reference substances, and specially prepared material which are relatively fragile and differ markedly in stability and quality during storage shall be retained only as long as the quality of the preparation affords evaluation. Specimens obtained from mutagenicity tests, specimens of soil, water-plants, and wet specimens of blood, urine, feces, biological fluids, do not need be retained after quality assurance verification.

(e) Master schedule sheet, copies of protocols, records of quality assurance inspections, summaries of training, experience and job description, and records and reports of the maintenance and calibration shall be retained for the length of time specified in 40 CFR 792.195(b).

Guide to Record Retention Requirements

in the Code of Federal Regulations

Revised as of January 1, 1989

Published by the Office of the Federal Register National Archives and Records Administration

LABOR DEPARTMENT

Occupational Safety and Health Administration

29 CFR

Part 1901 Contractors subject to Public Contracts Act (contracts with U.S. agencies or District of Columbia).

To keep a log, and an annual summary of occupational illnesses and accidents.

Retention period: 5 years following the end of the year to which they relate.

Part 1901 Contractors or subcontractors subject to Service Contract Act of 1965.

To keep a log, and an annual summary of occupational illnesses and accidents.

Retention period: 5 years following the end of the year to which they relate.

1904.2 Employers subject to the Occupational Safety and Health Act of 1970.

To maintain in each establishment a log and summary of all recordable occupational injuries and illnesses.

Retention period: 5 years.

1904.4 Employers subject to the Occupational Safety and Health Act of 1970.

At each establishment within 6 working days after receiving information that a recordable case has occurred, to maintain a supplementary record for each occupational injury or illness for that establishment.

Retention period: 5 years.

1904.6 Contractors or subcontractors subject to Service Contract Act of 1965. See 4.6.

1904.6 Employers subject to the Occupational Safety and Health Act of 1970. See 1904.2 and 1904.4.

1907.12 Accredited laboratories testing for safety specified products, devices, systems, materials, or installations. [Removed]

1910.20 Employers subject to record access rule on the preservation of employee exposure and medical records. [Revised]

To make available to employees and their designated representatives and the Assistant Secretary of Labor for OSHA (a) exposure; (b) medical records for examination and copying; and (c) analyses using exposure or medical records.

Retention period: Unless a specific OSHA standard provides a different time period: (a) 30 years, except for certain background data—1 year; (b) and (c) 30 years.

1910.38 Employers subject to fire prevention standards.

To maintain written records of emergency action plans.

Retention period: Not specified.

1910.68 Employers subject to manlift standards.

To maintain certification records of findings of manlift inspections.

Retention period: Not specified.

1910.95 Employers subject to occupational noise exposure standards.

(a) To maintain noise exposure measurement records.

Retention period: 2 years.

(b) To maintain audiometric test records.

Retention period: For the duration of the affected employee's employment.

1910.96 Employers subject to the ionizing radiation standard.

To maintain records of radiation exposure of all employees who are personally monitored.

Retention period: Indefinite.

-244-

1910.1005

Guide to Record Retention 1989

1910.1005 Employers subject to certain carcinogen standards. See 1910.1003.

1910.1006 Employers subject to certain carcinogen standards. See 1910.1003.

1910.1007 Employers subject to certain carcinogen standards. See 1910.1003.

1910.1008 Employers subject to certain carcinogen standards. See 1910.1003.

1910.1009 Employers subject to certain carcinogen standards. See 1910.1003.

1910.1010 Employers subject to certain carcinogen standards. See 1910.1003.

1910.1011 Employers subject to certain carcinogen standards. See 1910.1003.

1910.1012 Employers subject to certain carcinogen standards. See 1910.1003.

1910.1013 Employers subject to certain carcinogen standards. See 1910.1003.

1910.1014 Employers subject to certain carcinogen standards. See 1910.1003.

1910.1015 Employers subject to certain carcinogen standards. See 1910.1003.

1910.1016 Employers subject to certain carcinogen standards. See 1910.1003.

1910.1017 Employers subject to vinyl chloride standards.

To maintain (a) monitoring and measuring records; (b) authorized personnel rosters; and (c) medical records. Retention Period: (a) Not less than 30 years; (b) not specified; and (c) duration of employment plus 20 years, or 30 years, whichever is longer.

1910.1018 Employers subject to inorganic arsenic standard.

To maintain records of (a) employee exposure monitoring and (b) medical records.

Retention Period: 40 years or the duration of employment plus 20 years, whichever is longer.

1910.1025 Employers subject to lead standard.

To maintain records of (a) employee exposure monitoring; (b) medical records; and (c) medical removal records for employees removed from current exposure to lead.

Retention period: (a) and (b) 40 years or the duration of employment plus 20 years, whichever is longer; (c) duration of employee's employment.

1910.1028 Employers subject to benzene standards.

To maintain accurate employee exposure monitoring, measurement, and medical records.

Retention period: 30 years in accordance with 29 CFR 1910.20. All records shall be made available upon request to the Assistant Secretary and the Director for examination and copying.

1910.1029 Employers subject to coke oven emissions standard.

To maintain records of (a) all measurements taken to monitor employee exposure required by section cited, and (b) employee medical surveillance programs required by section cited.

Retention Period: At least 40 years or the duration of employment plus 20 years, whichever is longer.

1910.1043 Employers subject to cotton dust standard.

To maintain records of (a) sample exposure levels which would occur if the employee were not using a respirator and (b) medical records. Retention Period: At least 20 years.

1910.1044 Employers subject to 1, 2, dibromo-3-chloroprane (DBCP).

To maintain records of (a) employee exposure monitoring and (b) medical records.

Retention period: 40 years or the duration of employment plus 20 years, whichever is longer.

1910.1045 Employers subject to acrylonitrile (vinyl cyanide) standard.

To maintain records of (a) objective data relied upon in support of exemptions in the use of materials made from or containing acrylonitrile (AN); (b) employee exposure monitoring; and (c) medical records.

Retention period: (a) Duration of employer's reliance upon such objective data; (b) and (c) 40 years or the duration of employment plus 20 years, whichever is longer.

1910.1047 Employers subject to the ethylene oxide standards.

To maintain records of (a) all measurements taken to monitor employee exposure to EEO and (b) medical surveillance of employees.

Retention period: (a) 30 years and (b) for the duration of employment plus thirty years.

1910.1048 Employers subject to formaldehyde standards.

To maintain accurate employee medical surveillance records and accurate records for employees subject to negative pressure respirator fit testing. Retention period: (a) Exposure records and determinations shall be kept for at least 30 years; (b) Medical records shall be kept for the duration of employment plus 30 years; (c) Respirator fit testing records shall be kept until replaced by a more recent record.

1910.1100 Employers subject to the asbestos standards.

To maintain medical records and records of any personal or environmental monitoring required by cited section.

Retention period: 20 years.

1910.1200 Employers subject to hazard communication standards.

To maintain (a) a written hazard communication program for the workplace, including lists of hazardous chemicals present; labelling of containers of chemicals as well as of containers being shipped to other workplaces; (b) material safety data sheets that are received with incoming shipments of hazardous chemicals, and ensure that they are readily accessible to laboratory employees; and (c) copies of any material safety data sheets that are received with incoming shipments of the sealed containers of hazardous chemicals and shall ensure that the material safety data sheets are readily accessible during each work shift to employees when they are in their work areas. Material safety data sheets shall also be made readily available, upon request, to designated representatives and to the Assistant Secretary, in accordance with the requirements of 29 CFR 1910.20(e). The Director shall also be given access to material safety data sheets in the same manner.

1915.7 Employers subject to shipyard standards.

To maintain record or a copy of inspections and tests on file for each vessel.

Retention period: 3 months.

1915.12 Employers subject to explosive and other dangerous atmospheres standards.

To keep on file and make available for inspection a record of tests and inspections of atmosphere in the space to be entered.
Retention period: Not specified.

1915.97 Employers subject to health and sanitation standards.

To maintain an inspection form of hazardous materials to be used aboard vessels.

Retention period: 3 months.

1915.99 Employers subject to hazard communication standards.

See 1910.1200.

1915.113 Employers subject to shackles and hooks standards.

To maintain certification records of tests on all hooks for which no applicable manufacturer's recommendations are available before use.

Retention period: Not specified.

Guide to Record Retention 1989

1910.120 Employers engaged in the hazardous waste operations and emergency response operations under the Comprehensive Environmental Response, Compensation, and Liability Act of 1980, as amended (42 USC 9601 et. seq.).

To maintain records of the medical surveillance of (a) all employees who are or may be exposed to hazardous substances or health hazards at or above the established permissible exposure limits for these substances, without regard to the use of respirators, for 30 days or more a year; (b) all employees who wear a respirator; and (c) HAZMAT employees engaged in hazardous waste operations.

Retention period: As specified in 29 CFR 1910.20.

1910.134 Employers subject to respiratory protection standards.

To maintain records of inspection dates and findings for respirators maintained for emergency use.

Retention period: Not specified.

1910.156 Employers subject to fire protection standards.

To maintain written records of the fire brigade policy.

Retention period: Not specified.

1910.157 Employers subject to fire protection standards.

To maintain evidence of the required hydrostatic testing or portable fire extinguishers.

Retention period: The lesser of until hydrostatically retested at stated intervals or until taken out of service.

1910.159 Employers subject to fire protection standards.

To maintain central records of location, number of sprinklers and basis of design in lieu of signs at sprinkler valves.

Retention period: Not specified.

1910.160 Employers subject to fire protection standards.

To maintain records of the last semiannual checks of fixed extinguishing systems.

Retention period: Lesser of until the container is rechecked or its life.

1910.179 Employers subject to materials handling and storage standards.

To maintain monthly maintenance and test inspection reports concerning rated load test results and ropes idle for a month or more.

Retention period: Not specified.

1910.180 Employers subject to crawler locomotive and truck cranes.

To maintain monthly certification inspection reports and records on critical items such as brakes, crane hooks, and ropes.

Ropes shall be kept readily available.

Retention period: Not specified.

1910.181 Employers subject to derrick standards.

To maintain monthly written report readily available on inspections of all running and idle ropes.

Retention period: Not specified.

1910.184 Employers subject to industrial slings standards.

To maintain record of most recent month in which each alloy steel chain sling was thoroughly inspected and proof test certificates for each new, repaired or reconditioned sling; also, to attach permanent tag or mark, or keep a record in order to indicate date and nature of repairs to metal mesh slings.

Retention period: Not specified.

1910.217 Employers subject to mechanical power presses standard. [Revised]

To maintain records of the safety system installation certification and validation and the most recent recertification and revalidation as long as the press is in use. The records shall include the manufacture and model number of each component and subsystem, the calculations of the safety distance and the stopping time measurements. The most recent records shall be made available to OSHA upon request.

1910.218 Employers subject to forging machine standards.

To maintain certification records of the periodic and regular maintenance safety checks.

Retention period: Not specified.

Labor Department

1910.252 Employers subject to welding, cutting and brazing standards.

To maintain periodic certification records of maintenance inspections.

Retention period: Not specified.

1910.268 Employers, telecommunications.

To maintain on file certification records which include the identity of the person trained, the signature of the employer, or the person who conducted the training, and the date the training was completed.

Retention period: Duration of the employee's employment.

1910.272 Employers in grain handling facilities.

(a) To maintain on file hot work and entry into bins, silos, and tanks permits.

Retention period: Until completion of the hot work and entry operations.

(b) To maintain certification record of each equipment inspection containing the date of the inspection, the name of the person who performed the inspection and the serial number, or other identifier of the equipment inspected.

Retention period: Not specified.

1910.301 Employees subject to wiring design and protection standards.

(a) To maintain a written description of the assured equipment grounding conductor program.

(b) To maintain test records identifying each receptacle, cord set, and cord and plug connected equipment that passed the test and indicating the last date it was tested or the interval for which it was tested.

Retention period: (a) Not specified; (b) until replaced by a more current record.

1910.423 Employers subject to post-dive procedure standards.

To record and maintain information for each diving operation and for each dive outside the no-decompression limits, deeper than 100 fsw or using mixed gas.

Retention period: Not specified.

1910.440 Employers subject to commercial diving operations standard.

To maintain records of (a) occupational injuries and illnesses in accordance with 29 CFR Part 1904; (b) any diving-related injury or illness requiring hospitalization as required; (c) medical records; (d) safe practice manuals; (e) depth-time profiles; (f) diving records; (g) decompression procedure assessments; and (h) equipment inspections and testing records.

Retention period: (a) Medical records—5 years; (b) safe practices manual-current document only; (c) depth-time profile—until completion of recording of dive or decompression procedure assessment; (d) recording of dive—1 year except 5 years where decompression sickness, (e) decompression procedure assessment—5 years; (f) equipment inspections and testing records—current entry or until withdrawn from service; (g) records of hospitalization—5 years.

1910.1001 Employers subject to asbestos, tremolite, anthophyllite, or actinolite standards.

(a) To keep accurate records of all employees exposure measurements.

Retention period: 30 years.

(b) To maintain accurate record of objective data reasonably relied upon in support of exempted operations.

Retention period: Duration of employer's reliance upon objective data.

(c) To maintain accurate record for each employee subject to medical surveillance.

Retention period: Duration of employment plus thirty years.

(d) To maintain all employee training records.

Retention period: 1 year beyond last date of employment of the employee.

1910.1003 Employers subject to certain carcinogen standards.

To maintain records of medical examinations.

Retention period: Duration of employee's employment.

1910.1004 Employers subject to certain carcinogen standards.

See 1910.1003.

1915.172

1915.172 Employers of maritime employees.

To maintain certification records of the hydrostatic pressure tests on portable unfired pressure vessels.
Retention period: Not specified.

1917.23 Employers subject to marine terminal standards.

To maintain results of any tests on space which contains or has contained a hazardous atmosphere.
Retention period: 30 days.

1917.24 Employers of employees within marine terminals.

To maintain records of the dates, locations, and results of carbon monoxide concentration tests.
Retention period: 30 days.

1917.25 Employers subject to marine terminal standards.

To maintain test results to determine the atmospheric concentration of chemicals used to treat cargo.
Retention period: 30 days.

1917.28 Employers subject to hazard communication standards.

See 1910.1200.

1917.45 Employers of employees operating cranes and derricks.

To keep records of the monthly inspections of all functional components and accessible structural features of each crane or device.
Retention period: 6 months.

1917.117 Employers of employees within marine terminals.

To maintain manlifts inspection records.
Retention period: 1 year.

1918.61 Employers of maritime employees.

To maintain records of tests of strength of stevedoring gear.
Retention period: As long as such gear is in use.

1918.66 Employers subject to longshoring standards.

To maintain records of all hooks for which no applicable manufacturer's recommendations are available.
Retention period: Not specified.

Guide to Record Retention 1989

1918.90 Employers subject to hazard communication standards.

See 1910.1200.

1918.93 Employers of maritime employees under the Longshoremen's and Harbor Workers' Compensation Act.

To keep records of the dates, times, and locations of tests for carbon monoxide made when internal combustion engines exhaust into the hold or intermediate deck.
Retention period: 30 days after work is completed.

1919.10 Persons accredited for vessel cargo gear certification.

To maintain records of all work performed on gear certification, including tests, proof loads, and heat treatment; of the status of the certification of each vessel issued a register by such accredited person.
Retention period: Permanent.

1919.11 Persons accredited for vessel cargo gear certification.

See 1919.10.

1919.12 Operators or officers of vessels.

To keep vessel's register and certificates relating to cargo gear.
Retention period: 4 years after date of the latest entry except for nonrecurring test certificates concerning gear which is kept in use for a longer period, in which case certificates are retained as long as that gear is in use.

1925.3 Contractors or subcontractors subject to Service Contract Act of 1965.

See 4.6, and Part 1904.

1926.58 Employers subject to asbestos, tremolite, anthophyllite, and actinolite standards.

See 1910.1001.

1926.59 Employers subject to hazard communication standards.

See 1910.1200.

1926.251 Employers subject to rigging equipment for material handling standards.

To maintain records of tests on all hooks for which no applicable manu-

Labor Department

facturer's recommendations are available.
Retention period: Not specified.

1926.400 Employers subject to ground-fault protection standards.

To maintain records of testing of equipment required by section cited.
Retention period: Until replaced by a more current record.

1926.550 Employers subject to crane and derrick standards.

To maintain on file the most recent certification records which include date the crane items were inspected; the signature of the person who inspected the crane items; and a serial number or other identifier, for the crane inspected.
Retention period: Until a new certification is prepared.

1926.552 Employers subject to material hoists, personnel hoists and elevators standards.

To maintain on file the most recent certification records which include the date of the inspection and test of all functions and safety devices were performed; the signature of the person who performed the inspection and test; and a serial number, or other identifier, for the hoist that was inspected and tested.
Retention period: Not specified.

1926.800 Employers subject to tunnel and shaft standards.

To maintain records of all quantitative and qualitative air tests.
Retention period: Not specified.

1926.803 Employers subject to compressed air standards.

(a) A physician shall at all times maintain complete and full records of examinations made by him.
(b) To maintain permanent records of all identification badges issued.
(c) To maintain on file at the place where the work is in progress records of air tested in the workplace.
Retention period: Not specified.

1951.7

1926.850 Employers subject to demolition standards.

To maintain written evidence that a survey has been performed prior to demolition operations.
Retention period: Not specified.

1926.900 Employers subject to blasting and the use of explosives standards.

(a) To maintain inventories and use records of all explosives.
(b) To also maintain written descriptions of alternatives designed to prevent premature firing of electric blasting caps.
Retention period: (a) Not specified; (b) For the duration of the project.

1926.901 Employers subject to blaster qualifications standards.

To maintain on file written records of trucks inspected for the transportation of explosives underground.
Retention period: Not specified.

1926.903 Owners of trucks used for underground transportation of explosives.

To maintain on file the most recent certification records which include the date of the inspection; the signature of the person who performed the inspection; and a serial number or other identifier, of the truck inspected.
Retention period: Not specified.

1926.905 Employers subject to loading of explosives or blasting agents standards.

To maintain an accurate up-to-date records of explosives and accurate running inventory of all explosives and blasting agents stored on the operation.
Retention period: Not specified.

1950.11 State agencies receiving development and planning grants for occupational safety and health.

To maintain records consistent with pertinent instructions.
Retention period: Not specified.

1951.7 State agencies receiving grants implementing approved State plans in the occupational health and safety program.

To maintain financial records, supporting documents, statistical records,

-247-

1952.4

and all other records pertinent to the grant program.

Retention period: 3 years, or longer if audit findings not resolved; for non-expendable property, 3 years after final disposition. Microfilm copies may be substituted for the originals.

1952.4 Employers, except small employers as provided in 29 CFR 1904.21, subject to the Occupational Safety and Health Act of 1970.

To maintain records for each occupational injury and illness, including an annual summary, and also a supplemental record in detail according to OSHA Form 103 and such other records as specified in sections cited.

Retention period: 5 years following the end of the year to which they relate for records provided for in 29 CFR 1904.2, 1904.4, and 1904.5 (including forms OSHA No. 200 and its predecessor forms OSHA No. 100 and OSHA No. 102); 5 years for recordkeeping and reporting requirements in those States operating under State plans.

January 1, 1990 / Record Retention Supplement

Occupational Safety and Health Administration

29 CFR

1910.66 Building owners of all powered platform installations. [Added]

To maintain certification record which contains the date the work was performed, the signature of the person who performed the work, and an identifier for the equipment or installation which was tested or inspected. Records shall be kept readily available for review by the Assistant Secretary's representatives and by the employee.

1910.120 Employees engaged in the hazardous waste operations and emergency response operations under the Comprehensive Environmental Response, Compensation, and Liability Act of 1980, as amended (42 USC 9601 et. seq.). [Revised; effective March 6, 1990]

To maintain records or the medical surveillance of (a) all employees who are or may be exposed to hazardous substances or health hazards at or above the established permissible exposure limits for these substances, without regard to the use of respirators, for 30 days or more a year; (b) all employees who wear a respirator; and (c) HAZMAT employees engaged in hazardous waste operations.

Retention period: As specified in 29 CFR 1910.20.

1926.550 Employers subject to crane and derrick standards. [Amended]

To maintain on file the most recent certification records which include date the crane items were inspected; the signature of the person who inspected the crane items; and a serial number or other identifier, for the crane inspected.

Retention period: Until a new certification is prepared.

1926.652 Employers subject to the excavation occupational safety and health standards. [Added]

To maintain at the jobsite one copy of the tabulated data which identifies the registered professional engineer who approved the data and at least one copy of the design.

Retention period: During the construction of the protective system and after that time the data and design may be stored off the jobsite, but a copy shall be made available to the Secretary upon request.

1926.800 Employers subject to underground construction standards. [Revised]

(a) To maintain record of all air quality tests above ground at the workshop and make available these records to the Secretary upon request. The records shall include the location, date, time, substance and amount monitored.

Retention period: Until completion of the project.

(b) To maintain records of exposures to toxic substances in accordance with 29 CFR 1910.20.

About Government Institutes

Government Institutes, Inc. was founded in 1973 to provide continuing education and practical information for your professional development. Specializing in environmental and energy concerns, we recognize that you face unique challenges presented by the ever-increasing number of new laws and regulations and the rapid evolution of new technologies, methods and markets.

Our information and continuing education efforts include a Videotape Distribution Service, over 100 courses held nation-wide throughout the year, and over 150 publications, making us the world's largest publisher in these areas.

Government Institutes, Inc.
Rockville, MD (Washington, D.C.) 20850
(301) 251-9250

Other related books published by Government Institutes:

Environnmental Statutes, 1990 - All the major environmental laws incorporated into one convenient source. Hardcover/1,170 pages/Code 797; Softcover/1,170 pages/Code 796

Environmental Regulatory Glossary, 5th Edition - Records and standardizes more than 4,000 terms, abbreviations and acronyms, all compiled directly from the environmental statutes or the U.S. Code of Federal Regulations. Hardcover/456 pages/Code 798

Environmental Audits, 6th Edition - Details how to begin and manage a successful audit program for your facility. Use these checklists and sample procedures to identify your problem areas now and avoid costly compliance expenses later. Softcover/630 pages/Code 776

Fundamentals of Environmental Compliance Inspections - This new manual, developed by EPA for their inspector training course, allows you to get the same legal, technical and procedural insight into the basic underpinnings of all of EPA's compliance inspections. Softcover/300 pages/Code 782

Environmental Engineering Dictionary - Defines over 6,000 engineering terms used in pollution control technologies, monitoring, risk assessment, sampling and analysis, quality control and environmental engineering science. Softcover/630 pages/Code 786

Directory of Environmental Information Sources, 3rd Edition - Details hard-to-find Federal Government Resources; State Government Resources; Professional, Scientific, and Trade Organizations; Newsletters, Magazines, and Periodicals; and Databases. Softcover/322 pages/Code 221

Call the above number for our current book/video catalog and course schedule.

About Government Institutes

Government Institutes, Inc. was founded in 1973 to provide continuing education and practical information for your professional development. Specializing in environmental and energy concerns, we recognize that you face unique challenges presented by the ever-increasing number of new laws and regulations and the rapid evolution of new technologies, methods and markets.

Our information and continuing education efforts include a Videotape Distribution Service, over 100 courses held nation-wide throughout the year, and over 150 publications, making us the world's largest publisher in these areas.

Government Institutes, Inc.
Rockville, MD (Washington, D.C.) 20850
(301) 251-9250

Other related books published by Government Institutes:

RCRA Hazardous Wastes Handbook, 8th Edition - The Washington, D.C. law firm of Crowell & Moring gives you clear, concise answers to take you step-by-step through the maze of RCRA/Hazardous Wastes regulations. Includes the RCRA Statute. Softcover/490 pages/Code 778

OSHA Handbook, 2nd Edition - This practical non-legalese guide, written by a former OSHA official and now a practicing attorney, will help you meet your OSHA compliance concerns. Includes a new 76-page section "Suggested Forms, Practices and Programs for Use by Employers Who are Subject to OSHA Regulation." Softcover/400 pages/Code 769

Environmental Law Handbook, 10th Edition - The recognized authority in the field, provides practical and current information on all major environmental areas by eleven nationally-recognized legal experts. Hardcover/664 pages/Code 766

TSCA Handbook, 2nd Edition - The law firm of McKenna, Conner & Cuneo provides comprehensive look at your requirements under the Toxic Substances Control Act (TSCA). Includes a copy of the TSCA law, charts, tables, figures and multiple indexes. Softcover/490 pages/Code 791

Clean Water Handbook - Written by attorneys J. Gordon Arbuckle and Russell V. Randle of the Washington, D.C. law firm of Patton, Boggs & Blow, along with a team of other legal and technical experts -- offers a straightforward, non-legalese explanation of how the clean water laws and regulations affect business operations. Softcover/500 pages/Code 210

Call the above number for our current book/video catalog and course schedule.